GIS Data Sources

GIS Data Sources

Drew Decker

JOHN WILEY & SONS, INC.

New York • Chichester • Weinheim • Brisbane • Singapore • Toronto

Library of Congress Cataloging-in-Publication Data:

Decker, Drew.
    GIS data sources / Drew Decker.
        p. cm.
    Includes bibliographical references.
    ISBN 0-471-35505-4 (cloth : alk. paper)
        1. Geographic information systems—Information resources. I. Title.
G70.212.D44 2000
910'.285—dc21   00-033562

Printed in the United States of America.

10 9 8 7 6 5 4 3 2 1

■■■ CONTENTS

When searching for information about a particular subject today, it is easy to be overwhelmed by all the possible sources you are bound to encounter. And I am not just referring to the Internet! There are books, journals, conferences, courses, libraries, and professional knowledge and contacts that quickly provide you with more choices than you have bargained for. That can be a real problem. How do you sift through all that data out there to get the information you need? *Data* refers to material that is somewhat related to the subject of interest. *Information* refers to the elements that you really need and bother to stop and read.

In professions involving spatial data, this potential confusion is quite evident. Spatial data creation, storage, and usage have moved a long way beyond paper maps. Many professions—including core spatial-analysis fields, such as geology, cartography, and surveying, as well as other fields not immediately associated with mapping (such as environmental and urban planning, biology, transportation analysis, hydrology, and demographics)—use a tool now referred to as a geographic information system (GIS). A GIS allows you to organize and view spatial data and, more importantly, combine different spatial datasets to create new information that you did not have before.

Because all GIS practitioners require data, the intent of this book is to be useful to anyone involved in GIS. It is not a GIS primer and is not intended to help you operate a GIS. Instead, it is designed to help those who need GIS data to solve problems encountered at work, in the classroom, and in research. Many data sources are available, but there is little information about GIS data in general. That is what this book is all about: getting the most out of our existing GIS resources. The number of GIS datasets available, as well as their documentation, constantly increases, but there is still room for a reference book such as this.

My goal in writing this book is to help people locate the data they need. It's that simple. Data do exist, of course, and can be developed as needed, but without the opportunity to locate and exchange data, the unique benefits that GIS provides are restricted. Helping GIS users locate and get the most out of their GIS helps them, students of field, researchers, and benefits the field in general. I hope this guide helps you get to the information you need and that it will give you ideas on making more and better information available to others—a practice that will benefit all users of this versatile tool.

CHAPTER 1

Introduction

WHO IS THIS BOOK FOR?

Almost everyone involved in Geographic Information Systems (GIS) eventually needs data, so improving understanding and access to data can help everyone involved in GIS. This conclusion is rather broad so I will provide some specifics. This book is intended for people using GIS who have a fairly specific application that requires data not readily available. The data will have to be located and downloaded from a remote site. The users are comfortable with the basics of GIS and will be somewhat goal-oriented in that they know what information they want. People who fall into this group work for the government (local, regional, state or province, national, and international organizations), the private sector (consultants hired to address a specific problem), and academia (faculty locating lesson or research material and students attempting to complete the lesson).

The reason I cannot narrow down potential users any further than this reflects changes in the industry. GIS is spreading from an obscure computer map-generating technique typically housed in a little-understood "GIS shop" to a desktop tool used widely by students, instructors, and professionals in many fields. Describing the average GIS user may soon be like trying to describe the average word processing software user.

GIS data usage is very broad, so this work must also be broad. I can make some assumptions, however. This book will be most useful for those who

- Are familiar with the basics of GIS
- Currently use or are learning a common GIS software suite
- Require specific data to complete a project or otherwise better utilize the GIS

TABLE 1.1   Some Representative GIS Applications

Agricultural crop monitoring and modeling	Legislative district delineation/redistricting
Air pollution emissions inventory	Nonpoint source pollution analysis
Archeology surveys	Oil spill response and management
Census data analysis	Pavement management
Disaster relief assistance	Pipeline routing and permitting
Disease mapping	Population estimates
Drought planning	Property tax appraisal
Emergency Response (911)	Public lands and facilities management
Environmentally sensitive areas modeling	Timber tracking and permitting
Erosion studies	Transportation modeling and analysis
Flood analysis and modeling	Utility service area delineation
Groundwater modeling	Watershed delineation
Hazardous materials spill response	Wetlands resources mitigation
Land ownership mapping	Wildlife tracking

- Have the means to download data using the Internet or another computer network

For these reasons, I will not concentrate too much on the GIS basics, such as GIS operations (buffer, overlay, etc.), GIS implementation and planning, and hardware/software. I will touch on these subjects as needed, but it is the data that I'm interested in. This book will concentrate on common data sets that are being used today in GIS. The discussion on GIS data may swerve somewhat and examine older, perhaps obsolete data types as well as data sets that we will be seeing in the future.

GIS has many uses. Table 1.1 shows some common GIS applications, and the list by no means complete![1]

WHY WRITE A BOOK ON THIS SUBJECT?

With the advent of abundant GIS-related sources through the Internet, why put information in a book format? Why not put it solely on the Internet? For one thing, I want to create a more portable reference, something you can take with you from place to place. A book allows us to conveniently place material together in a more attractive and permanent method. Finally, people like reading from books more than screens. Reading out of a book is more pleasant for most people than staring at a monitor (access to high-quality laser printers notwithstanding!).

Beyond any arguments regarding format (book, HTML files, audiocassette, etc.), lots of information about GIS data is left unsaid. Data files of uncertain parentage are found throughout the Internet and scattered through most GIS shops. Because of this variety, increased usage of GIS, and frequent changes in data types I also want to discuss GIS data and how to use it. The proper application of GIS data has become complicated enough so that simple data lists are not sufficient. This book provides that GIS data discussion in a concise, plain English manner, explaining how to get the most out of GIS data.

That's it in a nutshell. There are lots of data out there but little information about the subject of the data. Specialized works that concentrate on a specific data type or GIS-related profession are fine, but I want to start with the data basics.

I work for a group that specializes in locating, storing, and distributing GIS, remote-sensing, and analog geographic information. No matter how many things are placed on the Internet or mailed to people, we always receive questions about the material. We all have questions about the data we need for our GIS projects. Are there better data out there? Should I make it myself? Where did the data come from? And, of course, How do I load the data and make it work? The contents of this book will help answer these questions.

WHAT WILL THIS BOOK ADDRESS?

This book has 10 chapters addressing GIS data types, sources, and problems and solutions, followed by a comprehensive reference listing. The chapters essentially address the questions people have (and we answer) about data. Chapter 2 describes the major data types that people use today—*The Basic GIS Data Groups*. Chapter 3 looks at what you need to decide before going on your hunt for data—*Defining Your Needs*. Chapter 4 asks, "What do you want it to look like?" It examines what the data will look like and how it will act when you get into your GIS—*Applying the Data: Envisioning a Finished Product*. Chapter 5 starts the fun stuff—*Locating the Data: The Sources*. I list common sources for data and types of data they offer. This chapter goes beyond lists, however, and states which data work and which do not. Chapter 5 does not just list sites, but says who has what and where they're going with their data.

Once we know where to look, it's time to see Chapter 6—*How to Obtain GIS Data*. This chapter details formats, media, compression, and various methods of downloading and moving data. Chapter 7 brings up the subject of *When Data Don't Exist—How can we make it?* Chapter 8 examines the *Keys to GIS Data Success*. Still making mistakes? Chapter 9 tells you the *Common Problems Encountered*. Finally, we look at what's coming down the road in Chapter 10—*Future Trends*.

The appendixes list a variety of GIS data sites that either serve data or link to other sites that do. Constant changes and additions to the Internet prevent a complete and comprehensive list, but these sites should help you get started. Appendix A lists the U.S. federal government sources. Appendix B has U.S. state government sources and Appendix C lists local government sources (primarily from the United States but from some other countries, too). Appendix D lists some good private company GIS data sources (both companies that do not create data, such as GIS software vendors, and consulting companies that create custom data as needed). Appendix E has foreign (public and private) sources. Appendix F lists those sources not particularly tied to one area or theme, such as universities and professional associations. This book is supported by an Internet site (www.gisdatasources.com) that has a more comprehensive source list.

You will notice something else in the book, too. In each chapter I identify a general rule that I think is notable. These GIS Data Laws are meant to be general rules of thumb describing trends or realities in GIS data usage. The first one can be found in the next section of this chapter and they are recapped in Chapter 10.

HOW CAN WE USE THE INFORMATION?

Before jumping into our discussion on GIS data sources, it would be wise to have a good idea of what we hope to glean from these data sources. In this book, the word *data* will be used frequently. I would first like to define it and place it in the proper context. *Webster's*[2] defines data as

> Datum \Da"tum\, n.; pl. Data. [L. See 2d Date.] 1. Something given or admitted; a fact or principle granted; that upon which an inference or an argument is based;—used chiefly in the plural.

Generally, the term *data* is used to describe basic facts. In earth science fields such facts may be the location of some phenomena, an elevation, or the relationship of one theme to another covering a given region. We note geographic data everyday—the location of a new building, for example. In geographic information systems we record these spatial phenomena in a digital manner. The phenomena, or data, if used intelligently, can be better understood as information and can help us form conclusions and solutions.

Figure 1.1 shows the classic information pyramid. This book will help you locate and better understand digital geographic data. You may not think there are enough data for your particular project, but when examined as a whole, there are more spatial data out there than you will likely ever see. You need to have basic *data* to load into your GIS. Helping you find and understand data is my job. Turning those data into useful *information* is your job (and that of your GIS). When properly interpreted and understood, this information increases your *knowledge*—not only in GIS but also in the phenomena you are studying, the region of interest, and the field where you will apply your conclusions. This knowledge helps you apply the information to solve your questions. When you have reached conclusions (the proper ones, I'll assume), you have arrived at the final step in the pyramid, *wisdom*, the ability to best utilize your knowledge to solve problems. The sizes of the steps in the pyramid are relative. Anyone remotely experienced with searching through Internet sites knows that data far outnumber information. Although this book talks about data, if I've done the right job, it should provide you with information leading to knowledge and even wisdom.

I've described this data-to-wisdom process very generally but the basic process takes place continuously as you gather outside facts to help answer a question. In GIS this is no different. We are awash in data, and, in keeping with the nature of data (not information), the geographical facts are hard to interpret. Cen-

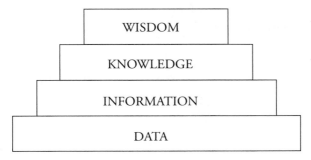

Figure 1.1 Data-to-wisdom processing pyramid.

sus information, archival satellite imagery, transportation computer-aided design (CAD) files, natural resource data layers—all are available. But the data are raw; they were not necessarily created to solve your problems. Instead, they were created by others to solve *their* problems. However, regardless of intent, there are still lots of uses for existing geographic data and they will likely be of some help to you, too. This leads to the first GIS data law:

GIS Data Law 1   The reason that many existing GIS data often cannot help you is that data are usually created to solve a specific problem and are not designed to be applied to a wide range of applications.

Fortunately, the GIS environment is changing too. More and more data are being created to serve as general base-mapping information that is more easily understood and available to many more users. The term *base map* will be used frequently in this text. Base map layers are intended to serve as starting points for more detailed mapping. They provide fundamental information relating to a particular data layer, or theme. The actual themes defined as base map layers are debatable but, by convention, most base map themes include the following:

- *Transportation layer.* This may include roads, railroads, trails, canals, and/or pipelines.
- *Land cover.* Regions are depicted by what land features are present (forest, wetlands) or by land usage (agriculture, urban, recreational).
- *Boundaries.* These include municipal, county, state (or provincial), and national boundaries. Often boundaries revealing specialized land holdings are included (parks, airports, military bases, and wildlife refuges and preserves).
- *Elevation.* This can include elevation contours or regularly spaced elevation points.
- *Hydrography.* Hydrography (often referred to a "hydro" for short) features include surface water features, such as streams and rivers, lakes, canals, and wetlands.

Other layers are sometimes referred to as base map layers, usually depending on the needs of a group developing maps for a particular region. Sometimes soil, land ownership, and utility maps are considered base map layers—it depends on the need for these maps and the budget for additional layers. Chapter 5 takes a closer look at base mapping and the groups that promote and fund these layers.

If we all use the same data as our starting point, we can more easily exchange the information, knowledge, and wisdom we derive from these data. Datasets are described in greater detail in Chapter 5. Data are becoming more easily understood because they serve general, not specific, needs. An example:

Simple—A digital contour map of Baffin Island in Canada

Complex—A digital contour map showing changes in extent and thickness of the Worthington Glacier, Alaska, from 1960 to 1990

You will find both kinds of data out there. General base map data such the Baffin Island contours are ultimately much more useful to GIS users. If you want the topography of Baffin Island, there it is. You can then use this general dataset in your GIS and combine it with other data to address any one of a number of issues. The complex example above is specialized information that is not likely to be useful to most GIS users, save Alaskan glaciologists. We will look more closely at these general digital base maps as well as metadata, which are summary files providing very detailed information on GIS data sets.

CHAPTER 2

The Basic GIS Data Groups

Many discussions of geographic data address various analog data types, including aerial photography, paper and stable-base material maps, and tabular data with spatial references. This book does not address these datasets but it is wise not to forget them. Many of the digital datasets we use originated from these sources. Know them, and you know their digital descendents better.

An excellent example of this is the ubiquitous 7.5-minute (1:24,000) U.S. Geological Survey (USGS) quad maps, which completely cover the continental United States with over 55,000 maps. These maps have been generated and updated continuously over the past 50 years. With the advent of GIS, much of the material on these maps is being updated and reproduced as digital data, through in-house development and by private contractors. Other federal agencies and state organizations acting as USGS partners are also converting the maps. The digital descendants of these maps are found in digital elevation models, digital line graphs, orthophotos, and numerous other products. The new digital data are available from many of the same sources that distributed the older paper maps. While anyone can find, download, and use these products, a good knowledge of how the original maps were created will save many questions later. That brings us to the following conclusion:

GIS Data Law 2   If you don't know the analog data, you don't know the whole story!

This chapter briefly describes many of the basic types of GIS data commonly used today. Most trace their beginnings to some form of analog data, including paper maps and aerial photos. The relationship between analog and digital products is important. Figure 2.1 is a cutaway illustration of an ana-

Figure 2.1. Split image showing a scanned USGS quad map on the right with its more recent, digital complement on the left. See any differences? The graphic on the right is from the 1:24,000 Lake Jackson quadrangle near Houston, Texas, compiled in 1977. The image to the left is the adjacent area shown on a 1:12,000 digital orthophoto created from a 1995 photograph. Sources are the USGS 7.5-minute Lake Jackson, Texas, quadrangle. Orthophoto image courtesy of Texas Natural Resources Information System.

log map flowing into a digital product. The right half of the image is a portion of a scanned 7.5-minute USGS quadrangle map. These maps have been serving the United States for decades. On the left is a continuation of the same USGS quad as shown in a digital orthophoto. I'll explain orthophotos later, but for now look at the differences. The orthophoto is corrected and spatially georeferenced (like a map), but it adds the detail of an aerial photograph. Orthophotos are now a standard product in the United States and in many other countries, too.

GIS COMPONENTS

The data discussed here, regardless of their differences, all serve as basic material for geographic information systems. A GIS is essentially composed of mapping software that combines spatial data (such as points, lines, and regions) describing features on the earth's surface with information further describing the spatial data.

There are numerous definitions of GIS, most of which describe to varying degrees the mapping and database capabilities of the software. The definition that I think is most accurate, however, is the following:

> A geographic information system is a geographic analysis tool that analyzes differences in multiple spatial data layers to create new spatial information not available by studying the data layers separately.[3]

This simply says that the real power in GIS is not in creating pretty maps or studying a data set definition repeatedly but in combining multiple data layers. For example, a digital elevation model (topographic surface) and a scanned aerial photograph of the same region can both provide information by themselves, but combine them and voila! Add a new view to the mix through GIS or image processing software and watch the information available to the user multiply. The sequence in Figures 2.2–2.4) shows the beauty of addition in GIS.

BASIC GIS DATA TERMINOLOGY

Know what raster and vector data are? Good, but it's important so I'll tell you again. Raster data are consistent, organized arrays of data. Data are composed of grid cells or intersections of the grid lines. An example of a grid cell is the values recorded for a cell (picture element, or pixel) in a satellite image or an orthophoto (see Figure 2.3). An example of data formed by a grid intersection is points in a digital elevation model (DEM) that record position (x, y or longitude, latitude*) and elevation (the z value). Figure 2.5 shows the varying resolutions seen in some raster data and the effect that has on the detail. In this case, the detail changes are evident in the hill-shade images.

---

*Note that I place longitude first, not latitude as is commonly spoken. x varies east/west, which is longitude, whereas y varies north/south, which is latitude. I think it is better to leave the common x,y pairing order alone than to adhere to the familiar "latitude/longitude."

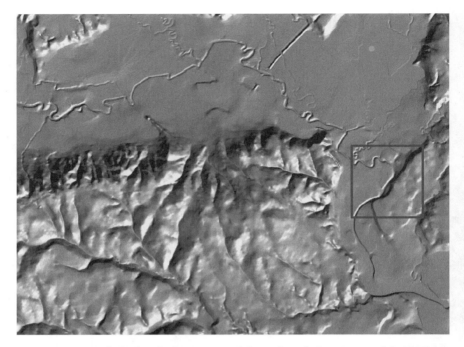

Figure 2.2.   Hill-shade relief map created from digital elevation models (DEMs). Digital elevation models are available for many parts of the world. The data, when combined with other GIS data, can produce very realistic three-dimensional images. The gray square shows an area where a digital orthophoto will be combined with the DEM. Sources are 1:24,000 USGS DEMs surrounding the Chireno North, Texas, quadrangle. Image courtesy of Texas Natural Resources Information System.

Vector data consist of familiar points, lines, and areas. Data points are just that—they represent a position to which further information can be attached. Because the position, such as longitude and latitude, is stored as an attribute to the point, it is often referred to as a zero-dimension element. Lines connect two end points, commonly referred to as nodes. Lines, of course, can be straight or curved. Curved lines have inflection points within, marking changes in direction. All lines can be measured in terms of length but not area. Therefore, they are referred to as one-dimensional elements. Areas (also called polygons) are enclosed two-dimensional regions that may contain or be intersected by points and lines. Areas, while obviously representing some sort of boundary, can have attributes further assigning a value to the entire region.

These are the basics. All GIS data are essentially one or the other. A good

Figure 2.3. This is the orthophoto covering the small area outlined in the previous figure. If you go back to the digital elevation model you can see slight depressions in the ground corresponding to the river and the lake in the orthophoto. The orthophoto is a 1:12,000 scale product (though it is printed at a different scale in this illustration). The area covered here is about 3.5 miles on a side. Image courtesy of Texas Natural Resources Information System.

way to remember and think of raster data is to imagine a continuous sheet, or layer, covering an area. A raster dataset is really a surface. Vector data, on the other hand, may continuously cover an area (i.e., groups of connected polygons) or may be intermittent, broken groupings of intersecting points, lines, and areas. All the data you get for your GIS will fall under one of these two types and the distinction between the two types is important. The software you use to an-

Figure 2.4. View of the orthophoto now overlying the digital elevation model. The image is viewed at an angle (oblique) to get the best three-dimensional effect. This is the power of GIS. The amount of information possible by combining GIS datasets far exceeds the information available by examining each layer separately. Image courtesy of Texas Natural Resources Information System.

alyze the data, the storage requirements, and the types of analyses performed on the data depend on whether the data are raster or vector.

You may be searching for vector, raster, or both types of data. It is quite common to use both types of data together in GIS. Raster data are commonly used as a background over which vector data are displayed. Figures 2.6 and 2.7 show simple linework (vectors) superimposed over a raster products (a hill shade and an orthophoto). It is also common practice to convert one form of data to another. Vector data are commonly derived from raster data in a process referred to as vectorization. Lines and polygons are created by connecting similar raster cells to create a continuous, unbroken line or a homogenous polygon. Vector data are not converted to raster data per se, but raster data are created when

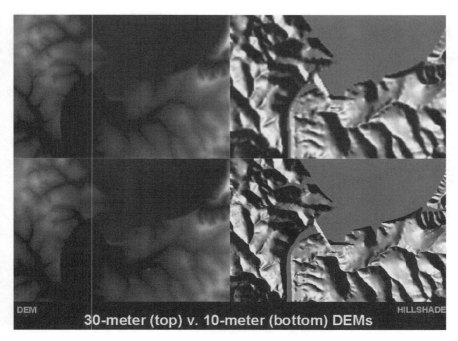

Figure 2.5. The images on the left are DEMs. DEMs are formed by a grid of regularly spaced points with elevation values. The grid points are 30 meters apart on the top image and 10 meters apart on the bottom image. This difference is rather difficult to detect by viewing the DEMs directly. If the data are processed further, however, the differences in detail are apparent. The hill shades on the right are artificial views formed by creating a surface between the grid points in the DEM and projecting light from an angle. See the differences in detail? They are particularly evident along the water's edge. Image courtesy of Forest Research Institute (Stephen F. Austin State University) and Texas Natural Resources Information System.

scanning vector features (points, lines, and areas) on a printed map or other graphic.

Searching for these data will take us to many varied sites. Many sources of data have both raster and vector data, but some specialize. Repositories specializing in satellite imagery, for example, would carry only raster data.

This ends the basic discussion of GIS and the two forms of GIS data. The next section examines different datasets commonly in use. Forms of data that are obsolete or not used (something can be commonly used and yet be obsolete and vice versa) may not be addressed here unless relevant to another dataset.

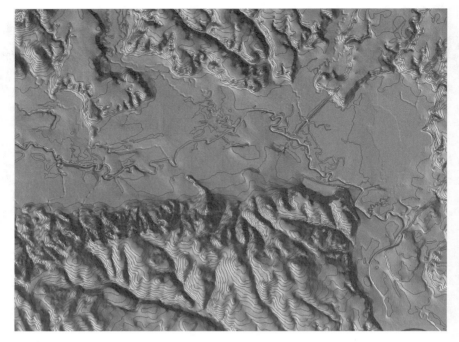

Figure 2.6.   Example of vector features (elevation contour lines) overlain on a raster background. Here, the raster background is the DEM-derived hill shade from Figure 2.2. Image courtesy of Forest Research Institute (Stephen F. Austin State University) and Texas Natural Resources Information System.

SOME NOTES ON SCALE AND ACCURACY

You will be better prepared to evaluate what data are best for you if you make a quick review of scale and format before you start searching for data. In geography, *scale* refers to the ratio of units measured on a map to the units measured on earth's surface. This is one of the most misunderstood rules in GIS and mapping in general. If the distance, for example, between two intersections on a map is 1 inch and the distance between the same intersections on the earth's surface (often referred to as the real-world distance) is 100 feet, the scale is 1:1,200 (1 inch on the map to 1,200 inches on the earth). The measuring unit type (inches, feet, meters) does not matter; just be sure to use the same unit in measuring both the map and the earth's surface.

The smaller the ratio of map units to real-world distance units, the larger the scale. Thus, a GIS dataset with a scale of 1:2,400 is considered a larger scale

Figure 2.7.   Second example of vector and raster data together. Transportation (roads) superimposed over an orthophoto. Notice how everything lines up. When data have equivalent accuracy and are projected in the same projection, you get the smooth vertical integration between GIS datasets seen here. Image courtesy of Texas Natural Resources Information System.

than one at 1:250,000. There is a general tendency for larger-scale datasets to contain more detail than smaller-scale data (see Figure 2.8), but this is not necessarily true. You can have a small-scale dataset covering a large area with a relatively high level of detail and a large-scale dataset covering a small area with very few features.

Another way to understand the significance of scale is its relationship with *accuracy*. The scale of data determines its general accuracy. Larger-scale data generally have a higher level of accuracy than smaller-scale data. Data accuracy refers to how close the features represented in the data are to their real-world positions. When you look at the coordinates of a feature in a GIS dataset, the dataset coordinates will be different than the actual coordinates on the ground. The measure of this difference is the accuracy of the GIS data.

How does one determine the accuracy of spatial data? On a map the best way is to look at the distance between map coordinates and true ground coordinates. Let's say you have found that features (road intersections, for example)

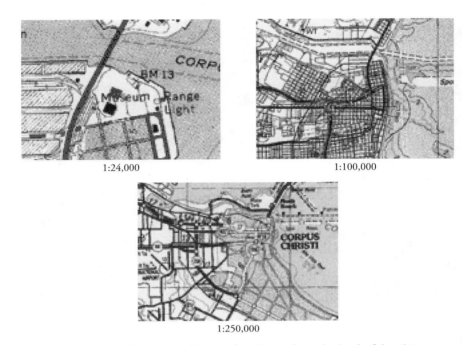

1:24,000   1:100,000

1:250,000

Figure 2.8.   Digital raster graphic samples. Notice how the level of detail increases with the scale. Images courtesy of USGS.

in your GIS dataset are off by 10 meters from the true location. How much is too much? The thought that something is off by "too much" relates to the scale that you wish to represent the data. The fact that the data are off is not unexpected—most spatial data are off to some degree.

Fortunately, there are some guidelines to help. The U.S. federal government, through the *National Map Accuracy Standards*, has equated accuracy and scale. A brief document can be found in many books and on a number of websites. One good site with the standard is rmmcweb.cr.usgs.gov/public/nmpstds/ nmas647.html. Let's see how the standard works and use the 10-meter discrepancy example above. The standard reads, in part, "For maps on publication scales larger than 1:20,000, not more than 10 percent of the points tested shall be in error by more than 1/30 inch, measured on the publication scale; for maps on publication scales of 1:20,000 or smaller, 1/50 inch. These limits of accuracy shall apply in all cases to positions of well-defined points only."[4]

Our 10-meter offset affects the scale to which we can say the data are accurate. The error amounts cited in the standard correspond to a certain distance based on scale. Let's see what scales can apply:

At 1:250,000 scale (larger regional applications), one-fiftieth (note that 1:250,000 is *smaller* than 1:20,000, so use 1/50 instead of 1/30) of an inch on the map covers about 417 ft (134 m) on the ground [(250,000/12)/50]. Our 10-meter difference is less.

At 1:24,000 scale (large scale suitable for detailed studies), one-fiftieth of an inch on the map covers about 40 feet (12 meters) on the ground [(24,000/12)/50]. The 10-meter difference is still less.

At 1:2,400 scale (highly detailed urban planning applications), one-thirtieth of an inch on the map covers only 6.7 feet (2 meters) on the ground [(2,400/12)/30]. Our 10-meter error exceeds this.

What does this tell us? The 10-meter error, though significant, does not preclude the data from being used. It can be used to accurately represent features to a scale slightly greater than 1:24,000. The calculation shows what happens when the wrong scale is chosen: The 10-meter error is far too great for the highly accurate requirements of 1:2,400 mapping.

This discussion was presented to help clarify scale. If you know this basic mapping concept and understand how accuracy and scale relate, then the rest is straightforward.

DATA SOURCES

The next two sections discuss the general types of GIS data found throughout the world today. I want this section to be somewhat generic so that it introduces data by its basic components (points, lines, and areas) and its common applications. I would rather people initially think about a dataset's applications and utility rather than the name of the dataset and other marketing information. I will be sparing with the marketing names of datasets and try to discuss data purpose. This will not always be the case—it is difficult to have a meaningful discussion of satellite datasets without using the word *Landsat*—but I will avoid name favoritism, such as, "Company X makes the best transportation (or whatever) data available."

RASTER DATA AND IMAGES

When we use the term *imagery,* different products come to mind. Orthophotos and satellite images are two of the most referred to raster images. Before discussing these products individually, we need to note a difference in the sources of these images. Both orthophotos and satellite data are called imagery but the orthophoto typically begins as a photographic product, which is scanned to become a digital image. The satellite image, on the other hand, is obtained electronically; there is no analog-to-digital conversion. All commercial satellite data today are wholly digital products. Whereas most orthophotos in use today began as photographs, an increasing amount of digital photography is becoming available. Airborne scanners are not new, but the digital photography that emulates typical black-and-white, natural color, and color infrared photography is. The next section on imagery addresses aerial photography and the popular orthophotos. The second section covers digital imagery (satellite images and direct digital photography).

Imagery—Photograph Based

Many sources of raster data in GIS begin as scanned photography. This generally refers to aerial photography aimed directly at the earth's surface below the platform. A line projected through the camera lens to the ground is perpendicular to the earth's surface. Photographs created when the angle of the line formed by the lens to the ground and the earth's surface is not 90 degrees (especially if that angle is substantially less than 90 degrees) are referred to as oblique photographs. Although they can be scanned and make great images and screen backgrounds, they are not considered acceptable as GIS data. Figure 2.9 shows a nice oblique aerial photograph.

Many scanned images are used in GIS both for illustrative and more serious, interpretive applications. For general imagery, where accuracy is not a major concern (and it may not be), a simple, scanned photograph may be acceptable. A simple, scanned aerial photograph may serve as a basic illustration or guide, but unless it is spatially accurate it has no real place in a GIS. All aerial photography is inherently inaccurate for a number of reasons, including parallax, fiducials, atmospheric interference, and platform instability. Distortion in aerial photography occurs primarily because the edges of the camera's view are farther from the lens than the center of the photograph.

Ortho Photographs

Aerial photographs used in GIS need to have the distortion removed. In essence, an ortho photograph is a corrected, digitized aerial photograph. Ortho photographs combine the detail of photographs with the properties of a

Figure 2.9. A beautiful image, but don't use it in a GIS! Notice the difference in scale in the oblique image—large scale near the camera and progressively smaller scales farther away. The near edge of the photograph is much closer to the camera lens than the farther edge. Image copyrighted by EarthData, International, 1999.

map. This means that distances and areas can be measured on the orthophoto, bearings can be taken, and the data can be combined with other GIS data. Orthophotos have been in existence a long time, although earlier orthophotos were primarily available only as analog products, such as a paper or film print. Image-processing advances have made it much easier to produce digital orthophotos that are stored as a computer image file. Figure 2.10 shows a closeup view of a standard USGS digital ortho quad (DOQ) with 1-meter pixels.

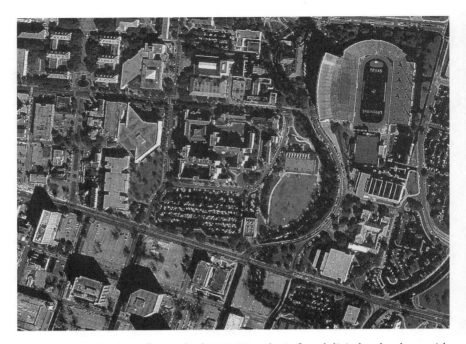

Figure 2.10. Portion of a standard 1:12,000 color infrared digital orthophoto with a 1-meter ground resolution. These products are becoming a standard GIS data set, not just in the United States but worldwide. Note that the image has been magnified to about 1:2,500 scale with no image distortion. Can you tell what university is pictured here? Image courtesy of Texas Natural Resources Information System.

How to Make an Ortho Photograph   Orthophotos are not hard to make, though the process can be laborious. A number of software packages are available that can guide you through the basic steps. Additional information besides your aerial photograph is still needed, so prepare to locate information about the camera that took the photograph and to get ground truth information (this might mean fieldwork—a good thing for getting away from the office!). The basic steps are as follows:

Step 1. Find a film positive of the aerial photograph you want to convert. The best way to envision a film positive is that it appears much like film negatives but without the pigments reversed. You can scan the normal paper print of the photograph, but the film provides a more stable base.

Step 2. The positive is then scanned at a high density (at least 800 dots per inch). The result is an image file ready to be corrected. These files are usually large. The typical 9 by 9-inch black-and-white photo will create a 50-megabyte file when scanned!

Step 3. Develop an aerotriangulation model that combines information on distortion introduced by the camera that produced the photograph, a digital elevation model to correct for terrain variation, and point location information to tie the photograph to the earth's surface. The model is essentially an equation that will take the scanned image and rebuild it pixel by pixel to eliminate distortion caused by the camera lens and platform attitude (i.e., tilt of the aircraft) and elevation changes.

Step 4. The new image is theoretically ready to use, although it is often further altered to change the format, projection, and datum to best match the other data you want to use with the orthophoto. The pixel size can be changed (this is called *resampling*) to change the image's detail and size.

The orthophoto recipe above is relatively simple but each step must be completed very carefully. After doing all this, what does an orthophoto look like? Orthophotos can be made from almost any aerial photograph (taken perpendicular to the earth's surface). Black-and-white, natural color, and color infrared film can all be used. Figures 2.7 and 2.10 are good examples of recent orthophotos. These are black-and-white orthophotos of downtown Austin, Texas area. The original color infrared image was scanned to produce a 1-meter pixel.

Digital Orthophoto Area   Orthophotos can be created to cover different sized areas with pixels of varying spatial resolution. Most orthophotos available for downloading today adhere to general areal standards set by the USGS. The standard USGS orthophoto covers an area 3.75 minutes on a side—one-quarter the area of the more common 7.5-minute USGS products. For this reason, USGS orthophotos are referred to as digital ortho quarter-quads (DOQQs*). This new size was chosen simply due to the size of orthophotos. One 3.75-minute USGS DOQQ contains approximately 50 megabytes of data for a black-and-white im-

---

*The term DOQQ is arbitrary in that it describes a portion of a specific quad size. The 3.75-minute quad upon which orthophotos are based is just as legitimate as the 7.5-minute quad. Thus, the term "digital ortho quad (or DOQ)" is more accurate and is less likely to cause confusion.

age and 150 megabytes for color infrared images. Thus, an orthophoto covering the standard USGS 7.5-minute quad would be 200 megabytes for black-and-white and 600 megabytes for color infrared. Images this size are too large to easily manipulate. A 600-megabyte image would be slow to load in any GIS or viewing software and cannot be easily downloaded or otherwise moved electronically, not to mention attempting various image manipulations. This is changing through better compression routines and faster transfer speeds (see Chapter 10) but 600-megabyte files are still difficult to work with.

Satellite Imagery

Satellite images form one of the most important sources of raster data for GIS and image processing software. As the name implies, imagery can be obtained from many earth-orbiting satellites providing data gathered by a variety of sensors. Satellite data are referred to as *imagery* instead of photography, because the vast majority of satellite data are collected through electronic sensors, not a film-based photographic process. A proper description of the many types of satellite data available is beyond the scope of this book; instead I briefly describe some of the more prevalent satellite data products. Sources for the imagery are provided here and in the appendixes.

Digital Elevation Models

Digital elevation models (DEMs) are one of the most versatile and useful GIS layers. These simple grid files contain locations (an x,y coordinate) with the elevation at that point (a z value) and support numerous topographical applications. The primary utility of these datasets comes from the fact that they represent a topographical surface (such as features on the earth) and show the interaction with physical landforms of all other data placed upon them. Other GIS datasets, both raster and vector, can be overlain on the DEM to provide a more realistic interaction between the data and the underlying topography. Raster data, such as DOQs and satellite imagery, when placed over a DEM, are shown in their true context as influenced by terrain. DEM data serve as the framework for many three-dimensional images and graphics.

In the United States, the term DEM generally refers to the well-known USGS product. USGS produces DEMs at a variety of scales from 1:24,000 (30 meters between elevation postings) to the worldwide Global 30-Arc-Second Elevation Data Set (GTOPO30) with elevation postings approximately 1 kilometer apart.

There is nothing restricting the creation or coverage of DEMs to the USGS. DEMs are available worldwide and are produced by many public and private entities. Companies make custom DEMs for their clients, and governments everywhere create DEMs for their territory. In the United States DEMs are available through the Global Land Information System (GLIS). GLIS is managed by the USGS Eros Data Center in Sioux Falls, South Dakota. Anyone needing access to the vast USGS digital data holdings will soon become familiar with EROS and the EarthExplorer system. The Internet sites are edcwww.cr.usgs.gov/ (EROS) and earthexplorer.usgs.gov (GLIS).

Another small-scale DEM-type dataset with global coverage is the Digital Terrain Elevation Data (DTED) available through the U. S. National Imagery and Mapping Agency (NIMA). DTED®0 Edition 2 is a terrain elevation dataset derived from NIMA VMAP Level 0 and is made available (within copyright restrictions) to the public at no charge through the Internet. Through 1997 and early 1998 NIMA completed near world wide coverage with DTED®0 Edition 2. Each DTED®0 data file contains elevation values for a 1-degree by 1-degree cell defined by whole-degree latitude and longitude boundaries. Each cell consists of a grid or matrix of elevation posts at 30-arc-second (approximately 3000-foot) intervals.[5] This is the same spacing as the USGS GTOPO30 data. DTED data are available through the NIMA website at www.nima.mil/geospatial/ geospatial.html.

Digital Raster Graphics

Digital raster graphics (DRGs) are a relatively recent dataset created by scanning existing paper maps and creating an image (with coordinates so it can be georeferenced) suitable for combining with other GIS data. This has been most commonly applied to USGS quad maps, particularly the 7.5-minute products, although many other paper maps could be scanned in the same manner. To create a typical USGS DRG a 7.5-minute USGS quad map is scanned by a large-format scanner at about 500 dots per inch (dpi). The image is then resampled to about 250 dpi to provide a more manageable image size while still maintaining legibility. Both USGS and their private-sector contractors are creating DRGs. Other vendors are creating customized products based on the basic DRGs. DRGs can be found at the USGS' EROS Data Center Internet Site (edcwww.cr.usgs.gov/webglis/). Figure 2.11 shows a full 7.5 minute DRG.

DRGs make excellent backdrops for vector data, providing additional context for the data you add. Because DRGs are available for most of the United

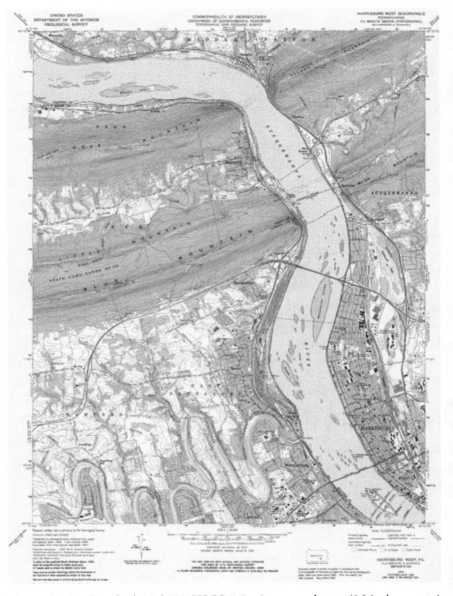

Figure 2.11. Standard 1:24,000 USGS 7.5-minute quad map (26 inches across) reproduced as a DRG file. The file is approximately 10 megabytes in size. The quad is Harrisburg West, Pennsylvania. Image courtesy of USGS.

States, you can add a DRG to vector data to better show the surroundings. In fact, this capability is one the best characteristics of raster data. Be it a satellite image, orthophoto, digital elevation model, or DRG, raster backgrounds show how vector data interact with the surrounding region.

Satellite Data

Digital data from earth-orbiting satellites provide a vast array of raster data suitable for use in a GIS. In fact, because satellites collect data continuously, it is difficult to keep track of the data available, let alone try to collect and store them. Because of the long history of satellite data and the variety of sensors in these satellites, I divide the satellites based on sensor types and then by individual satellite (or "bird" as they are frequently referred to). Although there are several thousand manmade satellites around earth, only a few series currently provide data suitable for GIS. These satellites are referred to as earth-observation satellites in that they collect images (from a variety of sensors) of the earth's surface and send these to receiving stations on the surface. There is one other important group of satellites that do not record imagery, but rather provide locational reference information concerning one's location on the planet. These locations are frequently used in GIS to record locations of surface features and georeference GIS datasets. The global positioning satellites (GPS) are the most complete satellite series currently used for locational information.

Earth-Observation Satellites

Landsat   The *Landsat* satellites started by NASA in 1972 are the best-known and most prolific series of earth-observation satellites. Continuously recording imagery from 1972 to today, *Landsat* provides up-to-date imagery over most of the world and also has a vast store of archival imagery suitable for many regional to small-scale GIS projects. Early *Landsat* satellites 1 through 3, provided multispectral raster data with 80-meter pixels. Their sensors recorded reflected energy in the green, red, and near-infrared portions of the spectrum.

Landsats 4 and *5* had an enhanced multispectral scanner called the Thematic Mapper (TM). The TM was a major advance in civil space-borne imaging because the scanner provided data in seven bands: three visible bands (blue, green, and red), a near infrared band, two midinfrared bands, and a thermal band. The thermal band did not detect reflected energy, but rather emitted energy, such as heat. The spectral resolution of the seven bands is shown in Table 2.1. The

**TABLE 2.1   Spectral Resolutions of the Seven Bands Provided by the
Landsat Thematic Mapper**

Band	Energy	Sensitivity (nm)
1	Blue	450–520
2	Green	520–600
3	Red	630–690
4	Near infrared	760–900
5	Midinfrared	1,550–1,750
6[a]	Thermal	10,400–12,500
7	Midinfrared	2,080–2,350

[a]Notice how the *Landsat* TM band numbers for bands 6 and 7 are reversed. TM was originally planned for six bands, including the thermal band. When the second midinfrared band was added, it was decided to call it band 7 instead of renaming the thermal band. Make sure you know what data you want when requesting *Landsat* bands 6 and 7!

spatial resolution for all the bands was 30 m, with the exception of the thermal band (band 6), which had a 120-m pixel. This spectral array provided additional increased resolution and much improved spectral range over the earlier *Landsats*. The blue band was useful for water penetration (although it was sensitive to atmospheric effects), and the middle infrared bands provided new geological and mineral information not previously available.

Landsat 7 is the latest satellite in the series. It was launched on 15 April 1999 and replaced *Landsat 6*, which failed to attain orbit in 1993.[6] *Landsat 7* employs an enhanced TM sensor system which includes a 15 meter panchromatic band.

As I stated earlier, it is far beyond the scope of this book to provide many details on all the types of GIS data available. Instead I want to point you in the right direction for locating data that will be beneficial, and to encourage you to get all the facts about the data and their sources possible. For example, see the http://geo.arc.nasa.gov/sge/landsat/lpchron.html Internet page for a great chronology of the *Landsat* satellites. Further information on the *Landsat* program can be found on the *Landsat 7* Homepage at landsat.gsfc.nasa.gov/. The USGS also has interesting information on *Landsat 7* at landsat7.usgs.gov/.

SPOT   The French Systeme Pour l'observation de la Terre (SPOT) earth observing satellites were launched starting in 1986. A second satellite was launched in 1990 and a third in 1993. The satellites were developed by the French Centre National d'Etudes Spatiales (CNES). Although Landsat data had been in use for number of years, the arrival of SPOT was well received due to the higher spatial resolution of the French satellites. The satellites have two sensors, a single band high-resolution panchromatic scanner, and a three band multispectral scanner which record, green, red, and near infrared energy. The 10m panchromatic and the 20m multispectral resolutions provided a notable improvement over *Landsat's* then highest resolution of 30m.* However, *Landsat* still collected a much wider spectral range of data.

The two types of SPOT images are often combined to produce merged imagery combining panchromatic clarity and the broader spectral range of the multispectral scanner. SPOT also offered off-nadir viewing capability—the ability to collect data not only directly beneath but also parallel to the track of the satellite. See www.spot.com for more information. The latest planned Spot satellite will offer 2.5m spatial resolution.

IKONOS   Much as SPOT improved upon the prevailing commercial satellite resolution, so did the IKONOS satellite when it was launched in 1999. Space Imaging, Inc. built and launched the satellite which offers a very detailed 1m spatial resolution from its panchromatic sensor and 4m multispectral data. The multispectral scanner collects four bands (blue, green, red, and near infrared) with the same spectral ranges as *Landsat* TM bands 1–4. IKONOS now combines the spatial detail at the common 1m DOQ level with the constant revisit properties of all commercial earth observing satellites. This product does not necessarily serve as a substitute for the DOQ but it clearly provides another alternative to people needing recent high-resolution imagery.

Like some other commercial satellites, IKONOS data are available through a network of retail partners as well as Space Imaging itself. See www.spaceimaging.com.

IRS   The Indian Remote Sensing Satellite (*IRS*) currently acquires data through several sensors and markets it commercially through a number of vendors. The latest satellite (*IRS-D1*) has a panchromatic band with 5.8-meter spatial resolution and a linear imaging self-scanning sensor (LISS) with spatial resolutions of 23 meters (visible and near infrared) and 70 meters (thermal infrared).[7] The satellite also has a wide-field sensor (WiFS) with 188-meter spatial resolution (red visible and near infrared). The higher-resolution *IRS* data complement imagery available from the *Landsat* and *SPOT* satellites and are available from sev-

---

*Note that 30m was the resolution of the *Landsat* TM data in 1986, when SPOT was launched. The enhanced thematic mapper scanner (ETM) on *Landsat* 7 has a 15m panchromatic resolution.

eral sources. The WiFS data are particularly useful for repeat coverage of large regions such as states and provinces. Like much satellite information, IRS data are available through a number of commercial resellers. Space Imaging, Inc. (www.spaceimaging.com) in the United States and Euromap in Germany (www.euromap.de/index.htm) are two well-known distributors of IRS data. Their Internet sites list facts about the satellite and data and imagery coverage; prices and order forms are also provided.

VECTOR DATA

Vector data form the features most people associate with GIS. The points, lines, and areas of vector data can describe almost any spatial feature you're interested in analyzing. Road networks, school districts, accident locations, addresses, rivers and streams—these features and many more are represented by vector data. Besides using existing vector data (one of the primary aims of this book), most GIS practitioners will end up making their own vector data. You can create vector data by digitizing off the computer screen (referred to as "heads-up" digitizing) or by digitizing a printed map on a digitizing board. Of course, you can also create raster data by simply scanning a map or a photograph, but these processes are more automated. Chapter 7 discusses creating GIS data in more detail. This section provides brief introductions to some of the more common GIS vector datasets. This list is by no means all-inclusive, but many of the vector characteristics described below are found in all vector data you may encounter in your data searches.

Digital Line Graphs

Digital line graphs* (DLGs) are digital representations of the vectors found on USGS quad maps. When you look at any standard USGS quadrangle map (regardless of scale), you will notice that all the data on the map can be grouped by general themes. For example, all the elevation contours on the map represent one data theme. The same can be said for the water features, transporta-

---

*The term *digital line graph* is not very accurate, having been derived from a time when digital vector data were relatively rare. The term causes confusion and is misleading; the term *graph* infers more of a raster product. DLGs are better described as simply the digital versions of the theme they represent, such as *digital transportation* or *digital hydrography* files.

tion features, manmade objects, and other feature groups common to these maps. Each theme is considered to be a separate DLG. These include transportation, hydrography (surface water features), hypsography (elevation contours), political boundaries, manmade features, cultural sites and miscellaneous transportation. To collect all the features you observe on the typical USGS quad digitally, you will need to get all the DLGs for that quad. You need to be aware of this added complexity when using DLGs.

DLGs are available in 1:24,000, 1:100,000, 1:250,000, and 1:1,000,000 scales—the same scales that USGS quad maps are printed. DLG availability varies. DLGs are not available for all quads, nor are all themes necessarily created should a quad have a DLG available. Most DLGs, though not all, are available in either DLG-optional or SDTS format. To increase the overall importability of DLGs, some files have been converted.

SSURGO

One recent addition to U.S. nationwide vector datasets has been the Soil Survey GIS (SSURGO) data. These datasets are digital versions of the common soil surveys read for many years in booklet form. The Natural Resource Conservation Service (NRCS, formerly known as the Soil Conservation Service) is a division of the U.S. Department of Agriculture and produces SSURGO data by counties. SSURGO data provide detailed soil information based on earlier county soil surveys and are updated when possible from the most recent aerial photography and orthophotos. Figure 2.12 provides a sample from a recent SSURGO dataset in Texas. SSURGO data vary widely in size, depending on the size of the county covered (ranging from 50 to over 5000 square miles) and the complexity of the soil environment. NRCS produces the SSURGO in Arc-Export format.

DCW

One of the earlier broadly used vector data sets is the small-scale (1:1,000,000) Digital Chart of the World, often abbreviated as DCW. The DCW is a basic digital dataset covering the entire planet and was derived from small-scale Operational Navigation Charts (ONC) developed by NIMA (formerly known as the Defense Mapping Agency). An excellent site with lots of helpful information on DCW as well as information on a value-added version of DCW is available through the Pennsylvania State University Library (www.maproom.psu.

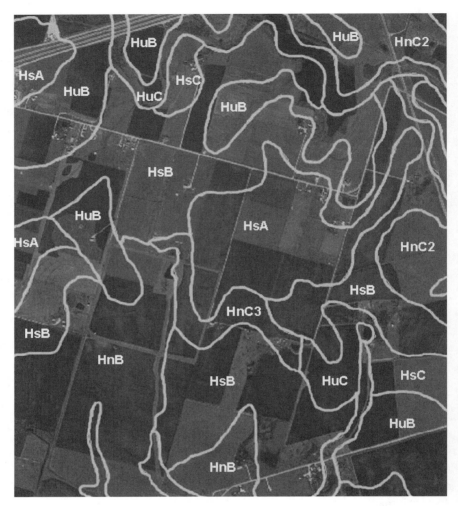

Figure 2.12. A relatively new GIS product—digital soil surveys. The latest soil surveys take data gathered through fieldwork, correct them to orthophotos, and attach attribute information further explaining each soil type. Image courtesy of Texas Natural Resources Information System.

edu/dcw/dcw_about.shtml). The data contain common thematic layers, such as transportation, hypsography (elevation contours), hydrography, railroads, populated areas, and some land cover. The Pennsylvania site provides a good description of the primary source data, the ONC:

Operational Navigation Chart (ONC) Product Specifications are designed and produced to support medium altitude en route navigation by dead reckoning, visual pilotage, celestial, radar, and other techniques. These charts are also widely used for mission planning/analysis, intelligence briefings, and the preparation of visual cockpit navigational display/navigational filmstrips. They provide a small-scale (little detail) translation of the cultural and terrain features for the pilots/navigators flying at medium (2,000 feet to 25,000 feet AGL) and low altitudes (500 feet to 2,000 feet AGL).[8]

CHAPTER 3

Defining Your Needs

While the primary aim of this book is to help you locate and access GIS data, you must know what to do with the data when you get it. I am not concerned with how the data are used (the GIS application) once it has been obtained, but you must know what the data will be used for before you download it. Chapter 2 introduced some of the common data types now used in GIS. To better use these data and to search for data effectively, you must have a clear idea of the role you expect the data to play. Before downloading data from the fastest website or running off to the nearest USGS office, work on a plan to ensure that the data that you do eventually get will work. The more you know about how the data will be used, the better position you are in to judge its value. This gives rise to the next GIS law:

GIS Data Law 3   Almost all GIS data have some value. Some data may require more manipulation but they can still make your GIS work better.

There are some exceptions to this rule, of course. I once found a GIS "file" showing electrical transmission lines across Texas. The apparent scale was very small, there was no projection, and I could not find out who made the file. A number of groups admitted handling the file and passing it along, but no one would admit to developing it. Denial was a wise move since the file couldn't be used in a GIS. Multiple attempts to reformat, load, and project the image ended in failure. Would you like to try? I could probably find the file and send it to you if you want to try working with it.

This illustrates the importance of data utility, which goes beyond mere data appearance. The transmission lines map looked interesting but was essentially useless in the GIS environment. This chapter addresses some factors to be aware of in planning for data usage.

WHAT DO YOU NEED THE DATA TO DO?

To answer this question it is a good idea to know where GIS data fit in the application. The more you know about the project and its goals, the better you can ascertain what data are best. This goes beyond the main intent of this work, locating GIS data, and is important no matter how you plan to obtain the data (site for existing data, in-house development, custom products, etc.). What you need to have is a summary of the project. Over the last five years standards have been developed to create summary information about GIS data, but knowing the project comes first. If you wanted to tell someone else about a project where you plan to apply GIS data, what facts would you list to create a summary? The summary information about data (see Chapter 5 for more information on metadata) follows a similar vein in listing answers to questions about a particular dataset.

Below are some things you should know about your project before you begin the search for GIS data. To help illustrate this, let's assume we are answering these questions for a hypothetical land-cover change study in a major urban area.

- *All specifications of the problem you want the GIS data to solve.* This point is critical. For land-cover detection and analysis, what information must the GIS data provide to be of assistance? What scale of data is best? How old can the data be? What attributes would you want any GIS transportation data to have (addresses, travel direction, road type, for example)?

- *Project timeline.* How long will it take to complete the project? When will you need to get the GIS data for the project? New GIS data are being created continuously. If the project will take some time to complete, there may be updates to the GIS data either in work or completed. The availability of GIS data change rapidly. If there are opportunities to import the latest GIS data into your project you should plan for this. Note that in this land-cover example new earth-observation satellites are becoming available. With the advent of the new 1-meter-resolution satellites, remotely sensed land-cover information sources in 1999 were notably changed from those in 1998.

- *Project participants.* What other groups will be participating in your project? Examining this can be one of the first steps in gathering GIS data. These groups could have data resources available or they could be charged with gathering or otherwise obtaining the data. In our example, if the city government is involved, then it may be best to assign the task of gathering the latest land-cover data to the city.

- *Project cost.* Once you know what funds the project requires you know how much money is available for data. Funding for data offers additional opportunity, such as being able to buy or lease the use of copyrighted data or to have data custom made for the project.

- *Study area the project encompasses.* The study area defines the region where you will be collecting data. This plays an important role in determining the types of data available, the temporal availability, and the scale. What part of the city do you need land-cover information for? Resolution and scale requirements can change, depending on whether data need to be gathered for the city center, incorporated area, suburbs, or the entire surrounding region.

- *Point in time that the project addresses (what is the date range?).* If your study covers a certain period in time (for example, land-use changes from 1990 to 2000) you will likely need data of different ages. Note that the GIS data available for one time period may not be the same type available at another date.

- *Will there be any follow-up work?* If the project is to become a series where data need to be gathered again in the future, this should be planned for now. Future trends in the types of data typically employed for this type of project should be studied so that the data gathered now will as compatible as possible with whatever is likely to be gathered in the future.

DATA PRODUCTION PROJECTS

The discussion above examined some factors applicable for projects where GIS data will be needed at some point in the project. The type of the project was not really important; the discussion revolved about how to plan for whatever GIS data the project would require. But what about projects where the primary intent is to create GIS data? The difference is in the intent of the project. In the above discussion, GIS data are a resource used as needed to solve a problem with a spatial component. There are additional factors to address when planning a project primarily to develop GIS data. These projects must (1) provide complete coverage of the area to be mapped, (2) be created at a suitable scale balancing sufficient detail with economy and dataset size, and (3) meet the needs of as many potential users as possible. The final point is critical and is usually the most difficult to meet.

Organization among project participants is important when building any spatial dataset. GIS coordination is needed because common data layers are easily and frequently duplicated by different entities. Government agencies, companies, universities, and other groups often create data when they need it even though there may be other data already in existence or being planned for by someone else. All project participants, even potential participants, must work together to plan common GIS datasets that they can all share. I see this happen with GIS planning at the state level. What if a state agency needs a statewide digital land-cover layer and the data currently offered through other agencies are not suitable? The most direct, though certainly not the cheapest, solution is to build a new land-cover layer. Legislatures began to notice funding requests for similar data-gathering projects by different agencies and this legislative awareness has helped support increased statewide coordination efforts.

Proper planning, though not too much—and I'll discuss this—is the key to building GIS databases. Building GIS databases is best described by breaking the process down into stages (we are assuming here that the data do not already exist, or if similar data do exist, that they do not meet our requirements):

1. Coordinate
2. Specify
3. Plan
4. Fund
5. Build
6. Distribute
7. Maintain

1. Coordinate. What data does your group need? Any group collaborating on GIS data production must be able to answer this question, and it is harder to do than you might expect. You must plan for a group, not just for a single entity. There are typically few resources devoted for intergroup planning efforts and most employees who can assist in planning for GIS datasets must remain loyal to the needs of their company or agency. Nevertheless, extensive planning is needed to avoid creating a sophisticated and expensive dataset that will ultimately not be useful or usable.

2. Specify. This process outlines the basic standard to which the dataset will conform. What existing standards will be used for the new dataset? What data format, projection, measuring units, datums, etc. are best suited for your mapping needs?

3. Plan. You need the backing of all participants and their intent to cooperate and basic dataset specifications before developing a plan. The plan outlines the entire project from start to finish and maps out the major phases in building the dataset. The plan lists the estimated budget, contractors, timeline, milestones, and key progress measures. The plan is the design for the entire project, but needs to have some flexibility built in for changes.

4. Fund. You must ensure that the project has adequate funds. Finding funding is a detailed discussion in itself and is made more complicated if the types of entities involved in the project are not specified. Funds can come from government sources (very common in much GIS data development) or a client, or can be provided up front by a group hoping to resell or lease the data.

5. Build. Finally, we reach the phase of the project where data are actually created. There are generally two avenues open for building data: They can be made in house by the group needing the data, or they can be created to your specifications by another group or company. Chapter 7 discusses this in greater detail. Projects can suffer from *too much* planning as well. Developing standards and plans to meet any contingency can stifle production. If the data meet your needs, build it. Sometimes standards develop behind (after) a good product has been produced. This way you get standards *and* data.

6. Distribute. How will the data be distributed to the project participants? This is not as easy as it may seem. Plan as much of the distribution process as possible. Getting extra copies of data made as part of the project is one thing; having to have additional copies made later is something else entirely. For example, a set of 1-meter DOQs covering a large area such as a state could fill thousands of CD-ROMs. Think now about distributing the data (format, media, compression, networks) rather than later. Another aspect of distribution is frequency. Will the data be distributed only once to project participants or will they be continuously distributed to a wider group? This question leads to the last part of the planning process.

7. Maintain. It is a good assumption (though not always true) that no one wants a white elephant dataset that is never supported. Filling a room with CD-ROMs, tapes, punch cards, etc. that cannot be accessed by the entities that funded the project (let alone other potential users) does no one any good. A useful dataset remains that way through constant maintenance. A process needs to be implemented for regular data updates and/or upgrades and to ensure that the data are available. This aim is best served by having a custodial entity oversee the maintenance. This is often a good role for government, because it can distribute data in the public domain. Groups that developed data using private funds are sometimes understandably hesitant to place data in the public domain. This can be done after a waiting period or by resampling the data to a lesser resolution before releasing them.

SUMMARY

Knowing what you want your GIS to do goes a long way to ensuring that you collect the right data. This is really the GIS version of doing the background planning and thinking that you would put into any project. It may seem as if I am going somewhat off track with this discussion on GIS projects. But the more familiar you are with your project, the easier it will be to find the right data and, better yet, the more time you can spend on the actual application itself without expending energy on data worries.

CHAPTER 4

Applying the Data—Envisioning a Finished Product

Chapter 3 stressed the importance of thoroughly understanding your GIS application (urban planning, transportation, land use, etc.) and the steps involved in common GIS project planning. This chapter takes this planning a step further and looks at some data concepts that can maximize the utility of the data in serving that application. What I want to do here is develop more of a strategy for getting the most from the data. We have discussed both different data types and how GIS can combine data to create new spatial data. When you apply these data to complete this application, you will have some form of end product. This product is more than just combined data; you need to think about how the product will be presented.

When you have gathered data and successfully solved your GIS application, what will you have left? Do you envision printed maps, diagrams, and reports? Will there be detailed custom GIS datasets designed specifically for your project? Will you be able to distribute the data you used to others for their own projects or will the data have been altered too much for other outside projects?

GIS Data Law 4   GIS data planning and manipulation is an ever-changing process. The amount of control GIS provides over your data is so great that you must continue to ascertain what products you want and how you want to present them throughout a project.

The GIS law above can be difficult to explain concisely, but I have found that the more knowledge you have about the final GIS product, the more options you will have to ensure that this product is feasible. This may seem obvious, but I have learned that it is difficult to take many things for granted in GIS planning. A number of factors must be considered:

- There are enough options and operations in GIS, CAD, and remote-sensing software that sometimes a new procedure can be employed to get a better product than was originally intended.
- All the factors that affect GIS operations (data availability, data types, standards, methodologies, and costs) are changing rapidly and will probably have some impact (for better or worse) on any sizable project, particularly one with a long timeline.
- Other technologies that GIS depends on (the Internet, computer speed and storage, and media) also change rapidly and could affect a project.
- Always be thinking of your customers. Their needs may change (if their needs do not change, their expectations might!) or perhaps you can think of ways to improve the product they're expecting.

The overriding concept is to accept that change is possible during the life of a GIS project. No one can predict all the changes possible, but you must be willing to accept this and be able to adapt to a fluid situation.

COMPONENTS OF FINISHED DATA

When you combine GIS data to solve some application, you will likely have one or more of the following products:

- A custom-made GIS dataset(s) made specifically for your application
- A group of individual GIS data layers that were combined to solve the application. These may be raw (unchanged) data from outside sources or layers that have been edited, or otherwise modified, to help solve the application
- Interim products
- Printed maps, reports, and other graphic information

Each of these products not only supports your application but also is presented in different ways and has its own level of utility beyond supporting the application. Let's look at some of these products and some considerations applicable to each.

Custom Datasets

Custom datasets are created strictly to support a GIS project or application. They are not intended for general distribution or for use as input to other projects. Custom datasets generally support a single conclusion and are often confined by strict positional, scientific, or temporal constraints. Examples might include

- An analysis of changing business patterns in downtown Seattle, Washington
- The location of traffic accidents in the Soho area of London between 1995 and 1999
- The locations of mammoth discoveries unearthed in Siberia on an archeological expedition
- Varying ice thickness in parts of Antarctica at the end of the Little Ice Age
- The properties in Monroe County, Florida, that are at most risk to category I hurricanes

Note how each example answers a specific question. These datasets help show the end result of research and support a particular hypothesis. The results serve a narrowly defined set of questions and are not broad in scope. They are intended for a specific audience—In the examples above it would be helpful to be familiar with the Little Ice Age or to know what a category I hurricane is. The data are still useful for historical comparisons or perhaps as input to other projects though their utility to outside projects in general tends to decrease with the complexity of the application.

Individual Data Layers

The second possible result from GIS projects has a wider range of uses. GIS is generally a tool of addition, with multiple layers of data combined to create new information (not merely data). These multiple layers can, of course, be made available for other uses. The utility of reusing data is obvious because whatever data you used as input data could be used again for other projects. When you have completed a GIS project, the input data may be discarded, archived, offered for distribution from your site, or, if updated, redistributed through your site or any other (such as the original host's).

Look at the examples in the Custom Datasets section above. Think of the input data that would be useful for these projects. In the first case, examining downtown business patterns, you might collect several datasets, including (1) Seattle road centerlines, (2) traffic count information, (3) land cover and land use, (4) building footprints, (5) retail sales records, and (6) utility information. It's also likely that you would retrieve different sets of these data covering a time series. In addition to retrieving and using these data you may also update some data with the latest information. Once the analysis is complete, what is to be done with the data?

- *Discard the data.* The data could be deleted and the space used for other work. This is especially feasible if the data are easy to obtain again. If there is any possibility that the data may be useful for a similar project then they should be kept.

- *Archive the data.* The Seattle input data could be easily stored on appropriate media such as CD-ROM, DVD, or tape. This is probably a wise long-term solution, because you cannot be certain of the availability of the same data in the future.

- *Distribute the data.* Why not offer the data to other users yourself? This is assuming the data were in the public domain to begin with, of course. If you feel the data would be useful to researchers with similar projects, this would be an opportunity to prolong the life of the data and help build ties to similar research.

- *Update the data and return it.* If you have had to update any of the input data for your project (e.g., collect and incorporate recent land-use changes into a land-use layer you found) you could provide the updated data to the source that you used. This is an excellent contribution to make, since consistent data updating is hard to implement. The original distributor of the data would appreciate this and it would help in getting data from that source in the future. The primary caveat to this, however, is that any data you provide back to the original source should be produced to their specifications or be otherwise easy to incorporate into their holdings. This is sometimes done in order to use certain proprietary data.

Interim Data

Some of the most important products created through GIS applications are *interim products.* Of all the resulting materials from GIS projects, interim products are the most easily overlooked, yet they can be very valuable for other projects. Interim products are datasets created during the processing of GIS data into their final form. Interim products are usually viewed solely as temporary files that are used as input for further processing or are discarded.

Numerous examples abound, such as the temporary files generated when applying algorithms to remotely sensed imagery or those created when revised (especially photo-revised) GIS data are being generated. These datasets can be valuable either as stand-alone products or as input to other applications beyond your own. The wise GIS practitioner will look to see what other uses these temporary data may serve. Two examples follow that illustrate common situations where interim data can be found.

Figure 4.1 shows one example of interim files being created through simple image-processing operations. Performing operations on large raster data alters the data and either creates a new file (as shown in Figure 4.1) or shows the file data on the screen that can be saved as a new file. Many of the image-processing operations result in interim files as large as the input file. Because satellite image and orthophoto files are so large, your automatic reaction is often to delete

INTERIM DATA IN IMAGE PROCESSING

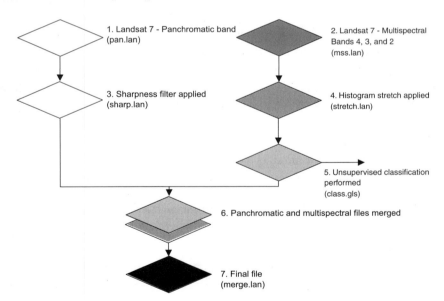

Figure 4.1.   Interim data resulting from common image processing operations. Some of you will notice that the file suffixes used here (.lan and .gis) are from older image processing software. There are other more current extensions, but these serve this illustration well.

or overwrite these files. If you are performing one operation, then you just save and use the result. However, many image-processing procedures require numerous steps, generating numerous interim files.

Looking further at Figure 4.1, note that for each step in any processing there is a file name in parentheses. This shows points where interim files are either created automatically (written to disk) or can be saved. There are several possibilities in handling these resulting files. Sometimes the new files (see the sharp.lan and stretch.lan examples in steps 2 and 4) can replace the original data for further processing (this is recommended only when you have the original data archived). Some interim data form a new product in their own right. This is seen where the classification produces the class.gis file (note that the file extension is now .gis because a thematic file has been produced in step 5).

Figure 4.2 shows another instance where interim files can be created: a production flow chart for creating hydrographic (surface water feature) vector files. The process has a number of different types of operations: data input, data manipulation, and quality assurance. As in image processing, the production of vector data also offers many interim products. A good example is step 2 (the box labeled "Color scanning, rectification, and extraction") where USGS quad maps are scanned so that line features can be extracted. When these maps are scanned, we get a raster file of the quad map that is similar to the USGS standard digital raster graphic product, or DRG (see the discussion in Chapter 2 on DRGs). In this particular production, the quad was scanned at 1000 dpi, thus producing an image with a far greater dot density than that of the standard USGS DRG's 250 dpi. Does it make sense to consider saving this scan file for other uses? It certainly does. Even though the scan file is not required once the vector lines have been extracted, the scan makes a fine supplement or replacement for the standard USGS product.

There are other examples I can cite where interim data have been useful:

- Photo-revised hydrography collected for digital soil surveys could be provided as additional input data for hydrography vector files (the same product shown in Figure 4.2).
- While extracting the hydrography lines from the scanned USGS quads, you could also extract other data layers (transportation, buildings, etc.) from the image as part of the same operation.
- Tagged vector contours (scanned and attributed contour lines) used to create digital elevation models (DEMs) could also be used as input for a totally different product, digital *elevation contours*.

HYDRO PRODUCTION FLOWLINE

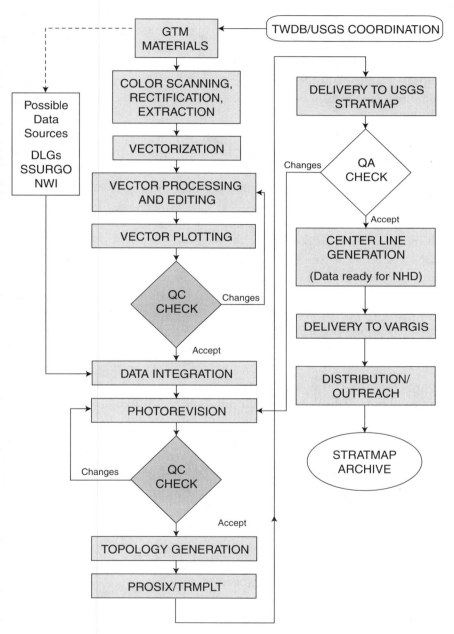

Figure 4.2   Interim data created through vector data production.

Interim data do not come without their costs, and in large projects the number of interim data to process could be staggering. The DRG example from Figure 4.2 would result in a 624-megabyte file to cover a 7.5-minute USGS quad. This is a huge file, but with storage costs decreasing, it could make sense to keep the file if it may be used in the future. With the wide range of GIS data and software permutations possible, I cannot recommend any set plan of action in addressing interim files, but want to stress that you need to be aware of other potential uses for these data.

Presentation

The discussions on custom data, individual data layers, and interim data dealt with the fate of digital data. What about GIS data presented in an analog format? Although the interpretation of GIS data primarily occurs within the computer, one of the most common ways to present findings is through a printed map. While data are stored in the computer, the spatial relationships between the features do not affect each other. When these features are printed, however, the intersections of features can cause problems, especially if several different data layers are printed together. Take, for example, a highway GIS layer that accurately depicts two separate roads only 20 feet apart. The GIS software will have no trouble keeping the roads separate, but the roads can appear to merge together when printed, depending on the scale of the plot and the road symbology. If this layer is printed at the common scale of 1:24,000, the two roads will be only 1/100 inch apart. With this distance it is likely not only that the eye cannot see that the roads are separate, but also that the map cannot be printed sharp enough to show this gap.

This is just one example of the factors that must be examined prior to producing a map from GIS data. There is a big difference between points, line, and polygons on a monitor and the same features printed on bright white paper in living color. Symbology, text placement and sizing, color lookup tables (for raster images), and legend design and placement are just some of the problems to watch for.

The best solution is to have a good knowledge of basic cartographic principles and to know how to apply them to GIS. A number of good reference books on this subject address how to prepare maps for applications, plotting methods, and general cartographic techniques. They include (and this list is by no means all-inclusive)

Bugayevskiy, L., and J. Snyder. *Map Projections: A Reference Manual.* London: Taylor & Francis, 1995.

MacEachren, A. *Some Truth With Maps: A Primer on Symbolization and Design.* Washington, DC: Association of American Geographers, 1994.

Monmonier, M. *How to Lie With Maps.* Chicago: University of Chicago Press, 1996.

Peterson, M. *Interactive and Animated Cartography.* Englewood Cliffs, NJ: Prentice Hall, 1995.

Robinson, A., J. Morrison, P. Muehrcke, A. Kimerling, and S. Guptill. *Elements of Cartography.* New York: Wiley, 1995.

Slocum, T. *Thematic Cartography and Visualization.* Englewood Cliffs, NJ: Prentice Hall, 1999.

I discuss the Internet and its impact on GIS data distribution numerous times in this book. Another bonus of the Internet is that finding the books above or any other reference material is just a quick Internet search away.

A final word on printing the nearby roads that I mentioned earlier. To ensure readability, features are often displaced hundreds of feet from their true location. This is fine when the end result is a small scale state highway map. But before moving features with a GIS for a printed map, it helps to have a good cartographic background.

DATA PACKAGING

One relatively new process in GIS data handling that can affect both data you need and data you distribute is data packaging. Data packaging is a loose term that describes integrating different data layers and providing data as a ready-to-use packaged set.

The world of GIS data distribution is becoming more than just that of making GIS data layers available and letting your customers pick through them as needed. One of the major changes in information technology today is convenience. Multipurpose communications, electronic banking, and the massive increase in electronic purchasing is based on making technology more convenient. Making a technology or function more convenient also makes it more useful and thus more utilized. The GIS data world, while large and getting larger, can make its data more receptive to more customers by packaging data to make it more convenient and easier to use for people who are not regular GIS users. Data packaging requires an appropriate region to cover, a suite of data to combine, vertical integration, and ease of access to the data.

Region Covered

The region covered is determined by the data intended for distribution and the customer. In the United States, many datasets are developed from USGS data. The current large-scale USGS data are available at the 1:12,000 and 1:24,000 scales. The obvious sources for these scales are the 3.75-minute and 7.5-minute USGS quadrangles. These are fine units for data development and storage but may not be the best for a specific audience. GIS resources (funding, staffing, and equipment) are greater for federal, state/provincial government, large corporations, and some universities; for smaller entities the resources are more limited. With these limitations comes the need for easier access to data. One of the first considerations in packaging is determining the data units: Transportation data may be assembled by the quad but in actual transportation applications, few users will analyze data by quad boundaries. Political boundaries are typically used instead. Thus, data confined by a city, county, state, or other district can be packaged. In addition to covering the appropriate area, properly packaged data cover only the area needed—there are no data to discard or otherwise store.

Data to Combine

Packaged data generally combine a number of data layers deemed appropriate to potential users. One possible combination is to provide base map, or framework, data (see Chapter 5 for more on framework). Base map data provide GIS data suitable for most general applications and provide a source from which new data can be extracted or added. Digital orthophoto quads (DOQs) are the best common base map layer, because they

- Serve as a good base map simply due to the amount of detail
- Permit most mapping functions (bearings, distances, areas, etc.)
- Allow additional data to be extracted (i.e., building footprints)
- Allow new data to be overlain or added (i.e., boundaries or utilities)

Other possible base map layers include hydrography (surface water features), digital elevation data (digital elevation models, contours, digital terrain models), survey and control, boundaries, transportation (including miscellaneous transportation such as railroads), and manmade features. There are no restrictions on what data may be packaged, though generally the data serve a common function, such as base mapping or planning. Packaged data should also be similar in scale (although all the layers do not have to have the same scale) and be of similar vintage. Once the data to package are collected, the data layers must be accurate with each other. This is referred to as vertical integration.

Vertical Integration

Let us look at a sample data integration instance. To package transportation and hydrographic data together we must be aware of the integration of the two layers. Given that the two data layers are at the same scale (1:24,000), cover the same region, are temporally consistent (data of the same vintage), and were collected using the same methods (from DOQs), what else needs to be considered? We need to look at the data together and see how they look relative to each other. Even though the data have the same accuracy, this simply means absolute accuracy. At 1:24,000 scale the road and water features cannot be displaced more than 12.2 m (40 ft) from their true locations. This is a good start, but the standards do not specify any direction of offset. This means that at any given point the road data can be off, say, 8 m in one direction while the water features might be off 6 m in the opposite direction for a total displacement of 14 m (46 ft) with respect to each other. This could cause a stream and road to lie atop each other, or water to be shown not beneath a bridge but flowing directly under the road! When we examine the data visually as a part of data integration, we can see where the problem is and fix it (perhaps by collecting more accurate data from GPS).

DATA RESAMPLING

Data resampling is a method commonly used to make a dataset more useful at other scales. You are not really creating new data, nor is resampling really an interim product. For these reasons it is in its own section. Resampling is the altering of the density of data elements in a dataset to make the data more suitable for other applications and for integration with data at other scales. Resampling as a concept can be applied to any data, but it is best explained with a raster data example.

If you begin with a raster dataset with a defined pixel size, the data can be resampled by combining pixels to create a coarser image. One-meter DOQs can become 10-meter DOQs; 10-meter satellite images can become 30-meter images. When the pixels are combined, detail is lost, so resampling is generally equated with degrading data. The 1-meter DOQ looks fine when printed at its design scale of 1:12,000. The same image resampled to 10-meter pixels would appear much too blurred and indistinct at 1:12,000; a better scale to print at would be 1:100,000. The same rule would affect the 10-meter satellite pixels. Scales appearing fine with the standard product would not be able to support the resampled product. Resampling employs the same principals with vector data. Points are removed from a line feature, for example, ultimately resulting in a less detailed, coarser product. While this is unacceptable when viewed at the original scale of the data, using the resampled data in smaller scale applications is fine.

Discussion thus far has centered on taking data out and degrading a product when resampling. Can the opposite occur? Can data be resampled to be used at larger scales? The answer is yes, but you must be able to combine the resampled dataset with other, higher scale data. Resampling a 4-meter pixel satellite image into 1-meter pixels by itself is meaningless. You are creating artificial detail from nothing. However, if that 4-meter satellite image is resampled to 1-meter pixels *and* the result is combined with another 1-meter dataset, then we could have a viable product.

Satellite imagery of varying resolutions has been merged for years. One effective combination was merging the 20-meter *SPOT* multispectral data with the 10-meter *SPOT* panchromatic data. A more recent version of this is the combination of 4-meter *IKONOS* multispectral data with 1-meter *IKONOS* panchromatic data (see Chapter 2 for more information on *SPOT* and *IKONOS*). In both cases, the larger image pixels are divided to match the size of the smaller, more detailed pixels. The more coarse multispectral pixels provide new information based on their infrared sensitivity and the panchromatic pixels provide the increased spatial resolution. Note that the comparative pixel sizes of these merged products do not differ too much. Merging a 4-meter product with a 1-meter product provides new information. Combining a 20-meter product with a 1-meter product, however, will not add as many data because the information provided by the larger pixels becomes more dilute the further it is subdivided.

A QUICK REVIEW OF THE BASICS

In conclusion, proper presentation and understanding of GIS data comes from a working knowledge of the basics. In my opinion, this means concentrating on the effects of position and spatial relationships between features. This knowledge should come before learning software, installing hardware, remembering terms, and learning how to digitize. Become familiar with the basics of spatial relationships—there are a number of excellent texts to help guide you. These books define terms, introduce data types, and discuss geometry and topology. It is the goal of these texts (and this book) to help you think spatially. Once you understand that locations are not simply where things are but their relationships to each other, the operations of GIS and the ability to express facts geographically will come naturally.

In the briefest terms possible, I consider scale, accuracy, and the knowledge of how accuracy influences scale to be keystones in properly applying geographic data. When these factors are understood, then you can understand the role they play when data are printed at a given scale. Data accuracy determines the scale to which data may be printed. This, in turn, provides a gauge of how "good" data appear when printed. For example, the common 1-meter DOQs are generally created with an accuracy supporting 1:12,000 representation (90 percent of the pixels must be within 10 meters of their true location). When the DOQ is printed on paper at 1:12,000, it looks right too. The detail is sharp; the pixels do not appear blocky. What if the DOQ accuracy exceeds the normal standard? If the pixels in the DOQ are generally within 5 meters of their true location, the DOQ could be used at greater scales. Using the formula provided in the National Map Accuracy Standards, a 1-meter DOQ with that accuracy could be printed at 1:6,000 scale, although just barely. Would the map look right? One-meter DOQs have been printed at a variety of scales for illustration and I can state that they would look fine. A 1-meter DOQ printed at a scale of 1:1,200, however, would not look right, regardless of the data accuracy, even if the pixels were within 1 meter of their true ground location. This scale is too large for a product with a 1-meter data element. The pixels appear blocky and surface detail becomes indistinct. You are essentially zoomed in too far.

Get some practice with how data are presented. Presentation becomes trickier with raster data, certain vector features (line thickness plays a role here), and when you add text. Keeping GIS data within their normal digital environment becomes second nature, but note that most data you will show to others wind up on paper. How you present these data can leave a lasting impression.

CHAPTER 5

Locating Data: The Sources

The material to this point has provided some background information you may need and how to best prepare for using that data when you receive it. This chapter is an introduction to the preparation of data sources you will encounter. Numerous entities support GIS data distribution, including government, the private sector, academia, and other groups, such as professional associations. The characteristics of these sources—types of data, staffing, download and other file transfer methods, relationships with other sources, and costs vary considerably.

Each source that holds and distributes GIS is unique. All sources have different histories, goals, and resources. There are also some basic differences between different groups of sources; for example, sites administered by the private sector will differ somewhat from U.S. federal sites. Regardless of source, the more you know about data the better decision you are able to make. Summary information about GIS data contribute greatly to determining its usefulness. So does organization of the GIS data and the themes available.

GIS Data Law 5   You may not be able to tell a book by its cover, but you can learn a lot about a GIS dataset from its associated metadata. Metadata are summary information (simplistically referred to as data about data) that provide a detailed description of the data. Accurate, up-to-date metadata are essential to efficient GIS data exchange.

FEDERAL GEOGRAPHIC DATA COMMITTEE

In 1990 the U.S. federal government created the Federal Geographic Data Committee (FGDC) to help coordinate and organize the creation, usage, and dissemination of geographic data at the national level. This is an interagency committee with representatives from 16 Cabinet-level and independent federal agencies.[9] The FGDC has been instrumental in organizing the exchange of vast amounts of geographic data both among federal agencies and between these agencies and the rest of the GIS community. This body of data and the role it serves has been labeled the National Spatial Data Infrastructure (NSDI) by the FGDC. The role of a coordinated National Spatial Data Infrastructure is "to support public and private sector applications of geospatial data in such areas as transportation, community development, agriculture, emergency response, environmental management and information technology."[10] One major accomplishment of the FGDC's support of this geospatial infrastructure has been to establish standards and protocols for sites dedicated to serving geographic information. The establishment of the data "clearinghouse" concept can best summarize this.

CLEARINGHOUSE

The Clearinghouse concept deals with the real meat of data distribution. Clearinghouse has been promoted by the FGDC as an organized method of distributing GIS data. Sites distributing GIS data are nothing new. Virtually every type of entity one can imagine has been distributed data for GIS. Government (all levels, from local through international), universities, companies, and professional associations have all delved in distributing data. All intentions have been good, but results have varied: The types of data being distributed were different, the data lineage was often poorly documented or unknown, and information on projection and datums could be misleading, if described at all.

All these concerns led to data distribution being much harder than it had to be. The FGDC, through the NSDI, led an effort to devise a more "official" method of distributing data. A data clearinghouse would be a site dedicated to data distribution that followed the basic FGDC guidelines for data organization and metadata referencing. Data available through an NSDI-sanctioned clearinghouse would need to be properly referenced, and organization of the data types and the features they covered must be consistent. The Clearinghouse section at the FGDC Internet site has full details. An NSDI Clearinghouse site is a site dedicated to data distribution that followed the basic FGDC guidelines for data organization and metadata referencing.

BASE MAPS AND THE FRAMEWORK CONCEPT

One of the concepts FGDC addresses is the notion of base mapping. The term is used frequently in GIS data development today and it is easy to get confused. What do we refer to when we describe a GIS dataset as a base map layer? I describe a base map as follows:

> A base map is a theme that provides essential information on common land features upon which mapping applications may be performed and from which more specialized data may be derived. Typical base mapping themes include features common to any given region such as transportation, elevation, hydrography, land cover, and boundaries. The broad range of features collected in base maps means that multiple groups can share the same data. There are few set rules on what can be a base map layer or what the scale or the level of detail should be. Determination of these characteristics depends on the needs of the organization developing the data for further use.

The proper production of base map layers is important to the FGDC. The FGDC coordinates the geographic data development of many large federal agencies, each with their own needs and budget limitations. If we examine what geographic features are common to most of the needs of these agencies, we have begun to develop a logical set of base-mapping layers. Some problems found with data distribution sites were listed in the previous section. One concern was with the actual types of data being distributed. As practitioners of GIS, we know that virtually any theme may be captured through GIS. But, are some themes more appropriate than others? Add the array of features we can represent in GIS to the variety of scales and it can get quite confusing. The framework concept helps to define two areas: (1) What are the best base themes to collect, and (2) at what scale are those data collected?

Other terms are used interchangeably to describe these features, such as foundation data and the FGDC's framework. For the sake of continuity, I use term framework data in this chapter because it ties in well with the FGDC focus of this chapter. In other areas of the book the more generic term base map is more appropriate.

FGDC's framework concept takes the base map idea one step further. Framework data contain not only base-mapping features, but also the "best available" data. For example, a transportation GIS dataset for a given region may be recognized as a base map layer, but is it the best data available? The best available data position states that just having the data may not be enough. For example, you can locate transportation data for a given city from numerous sites, but these sites may not have the latest, most detailed, or best-maintained data.

Why are these features thought of as framework features and not other common GIS phenomena, such as addresses, census data, weather information, and traffic data? Because framework features are tied to a specific location on the earth (unlike changing features such as weather), they can be easily classified and defined (unlike addressing schemes, which can vary), and they are tied to physical, unmistakable topographical features (unlike census data, which is tied to artificial boundaries).

Framework helps narrow the wide range of geographic phenomena and provides a good starting point for further GIS applications. No one wants to restrict the range of applications in GIS. Providing framework data gives GIS users a common ground to start from. Let's agree on common base map layers* first so that GIS data can be more easily exchanged later. For example, let's say one region defines transportation and land cover as its base map layers. This sounds reasonable, right? But the adjoining region has defined school districts and utility districts as its base layers. Where's the commonality here? Any attempt to use GIS for applications between these regions with these layers will be stymied.

Framework also refers to scales of base map GIS layers. In the United States, there are several common scales to which data are accurate. We commonly encounter scales of 1:24,000, 1:100,000, 1:1,000,000, and so on. Why do we see these scales so frequently in GIS? We can use data accurate to many varying degrees and essentially have data fit for mapping at any scale we desire. The reason we see these scales is that so many GIS data have been derived from older printed maps (see GIS Data Law 2). However, it is true that GIS can use data accurate to many scales beyond these older common scales. The orthophotos, now being thought of as the new digital base map layer, are produced at 1:12,000 scale—no prior national coverage USGS maps had that scale. Local governments are creating data at 1:6,000 and better.

Framework recognizes this, and due to this scale variance, simply seeks to

---

*Note that I use the terms "layers," "theme," and "features" to refer to essentially the same thing, namely, what it is we are trying to represent with our map. A GIS data file with roads in it will be called a road layer, road features, or even a road theme (infrequently used). These terms all mean the same thing. Of these names, the most technically accurate descriptor is probably "feature." The term "layer" is very common; though it is based on terminology ascribed to data handling by software, that has not stopped many people from using the word frequently, including me.

find the best scale (most accurate) data for a given region. It doesn't say we have to collect data at 1:24,000 or 1:250,000 or anything. Framework wants to link the data covering adjacent regions, and if data have different scales, that is to be expected. One way to represent this is by looking at different data scales as a surface. We can imagine a hypothetical surface formed by different scales. In this sample region, see how scale varies from point to point. This is reality. Imagine that a rectangular region is the province of a country (no particular one). The highest points of the surface represent the highest scales—most likely found in major cities. Areas with midlevel scales may represent a small regional government that needs some detail, but not as much as needed to record features of a city, such as sidewalks, signs, and manhole covers. The lowest areas of the surface are, of course, the rural areas with little infrastructure and there is no need to record the more detailed features. You might want to map land cover in rural areas at 1:12,000, but would almost certainly not need the positional accuracy or feature level of detail at 1:1,200.

The FGDC Internet site further describes framework as "a collaborative effort to create a widely available source of basic geographic data. It provides the most common data themes geographic data users need, as well as an environment to support the development and use of these data. The framework's key aspects are seven themes of digital geographic data that are commonly used: procedures, technology, and guidelines that provide for integration, sharing, and use of these data; and institutional relationships and business practices that encourage the maintenance and use of data." Table 5.1 below lists the themes considered by the FGDC to be framework layers.

TABLE 5.1   List of FGDC Framework Data Layers

Theme	Features Included
Elevation and bathymetry	Elevation points and contours. Bathymetry refers to elevations (depths) beneath a water body
Hydrography	Surface (not underground, such as aquifers) water features, including canals, streams, and lakes
Geodetic control	Survey and marker points
Cadastral	Land parcel ownership and boundaries
Transportation	Roads, railroads, trails, canals, and can include miscellaneous routes, such as pipelines
Governmental units	Political boundaries and jurisdictions
Digital orthoimagery	Digital aerial photography

In summation, the framework section of the FGDC Internet site (www.fgdc.gov/framework/overview.html) reiterates the aims of the framework concept very well:

> The framework represents "data you can trust"—the best available data for an area, certified, standardized, and described according to a common standard. It provides a foundation on which organizations can build by adding their own detail and compiling other data sets.

METADATA

Metadata are best described as "data about data." Metadata essentially provide a summary with detailed information pertaining to a particular GIS dataset. It has been the goal of the FGDC in the United States and other organizations abroad to promote the creation and usage of metadata as part of the routine in using GIS datasets. There has been much discussion about metadata since the first standardized content was released by FGDC in 1994, with minor revisions in 1998. Metadata are undergoing further revisions, with the FGDC working "to harmonize the FGDC Metadata Standard (FGDC-STD-001-1998) with the International Organization for Standardization (ISO) Technical Committee (TC) 211 Metadata Standard 5046-15."[11]

What is a metadata file? First, remember that the metadata standard is a *content* standard. This means that the standard defines the questions to be answered within the metadata file. The metadata file itself is simply the answers to these questions presented in an organized fashion conforming to the order in which answers are provided and the spacing between answers so they can be read by programs designed to elicit metadata responses. That's a mouthful, but an easy way to envision the metadata file is to imagine a word processing document containing facts about the GIS dataset with a certain spacing and organization within the file dictating where the answers go.

Data can be entered into a metadata file through several methods. The data can simply be entered into a word processing file template. If you have filled out standardized Word or WordPerfect forms before, you should have an idea what these templates look like. You can also enter metadata through a number of software tools that provide forms for you to enter information or otherwise prompt you for answers. The FGDC lists metadata entry software sources at www.fgdc.gov/metadata/toollist/metatool.html.

Metadata Content

OK, but what's in the metadata file? The metadata contents are the centerpoint to the whole metadata concept. The FGDC has specified a long list of questions to answer to best describe any GIS dataset. Think about what people would want to know about a dataset you're working on now. They will probably want to know the

- Scale (e.g., 1:24,000)
- Author or creator (whoever made the dataset)
- Time (when was it finished)
- Subject or theme (highways, census blocks, flood likelihood, etc.)
- Projection (Lambert conformal, Albers, custom, etc.)

Well, the FGDC provides all these questions and more in their content standards. The full FGDC metadata content contains some 200 questions, though not all are required. The current metadata content standard is FGDC-STD-001-1998, *Content Standards for Digital Geospatial Metadata* (CSDGM—revised June 1998), Federal Geographic Data Committee, Washington, DC.

When devising these questions the FGDC had to plan for virtually any question someone might have about a dataset at a later date. This approach is taken with all sorts of categorization. What information might you want to know about a particular song? You may ask questions about the artist, music genre, album, type of recording (studio, demo, live), instrumentation (acoustic or electric), media format (record, tape, CD), etc. Similar questions about artworks and literature can be asked. The FGDC looked at the tenants of classification for all these items to ensure that all possible questions were asked.[12] One interesting way to look at metadata is to think about the labels on packages of food. These labels are everywhere and list calories, nutritional content, ingredients, and so on. The FGDC and other metadata instructors commonly use this example in explaining the need for metadata.[13]

Not all the questions in the FGDC content need to be answered. Questions are categorized as "Mandatory," "Mandatory if Applicable," and "Optional." Mandatory questions address the basic facts about the dataset, such as who created it, organization, date, and projection. "Mandatory if Applicable" questions apply to facts that depend on the dataset type and parameters associated with a particular projection. A raster GIS dataset would have its own group of "Mandatory if Applicable" questions, for example. Optional fields allow the metadata author room for additional details on aspects of the dataset.

While metadata files are valuable in helping future users gain the most from the material, they can be difficult to complete. A completed metadata file (with all questions answered) is generally referred to as *FGDC compliant metadata.* Metadata files that cannot be completed are termed, of course, *noncompliant.* When you have completed entering the metadata, the file is run through a check program called a *metadata parser* that ensures that the right fields have been filled in and the results meet the organization specified in the CSDGM. Why run your metadata file through the parser? This is done so that the metadata file is standardized and can be much more easily exchanged with and read by other people not involved with your efforts. Just as USGS files are standardized so that all users can read them, the same logic follows for metadata.

The FGDC has a number of utilities, publications, and slide shows to assist with metadata implementation. The mp (metadata parser) utility was developed by the FGDC and is available from a number of sites. The latest version of mp is Version 2.4.35, released on 23 November 1999. The Wisconsin State Cartographer's Office has an excellent Metadata Toolbox site that carries mp and other utilities. The site is located at badger.state.wi.us/agencies/wlib/sco/metatool/. This is not the sole site for metadata-related utilities, but it is well organized and is a good resource for someone starting metadata maintenance.

Running a metadata file through mp tells if your metadata meet the CSDGM content standard. Metadata file construction depends on a strict set of indentation rules to separate different content categories. If you put in a value not recognized in the standard or one out of the range of possible responses, mp will find it. If the hierarchical ordering required by CSDGM is not followed, likewise, mp will alert you. The mp utility works best for files that are essentially in the correct CSDGM format and are not expected to deviate much from the standard. If you have entered metadata information and feel that the material is not really ready for mp, you can run through a utility called cns (Chew and Spit)—Version 2.3.3, which is a preprocessor for mp. As its name implies, cns will "chew" through raw metadata information and produce an output more compatible with mp. This utility, as well as many others, is also available from the Wisconsin State Cartographers Office site.

Metadata files are best handled when they are completed in conjunction with a new GIS dataset. If not, going back to answer all the questions the FGDC content demands can be difficult. This problem is even more apparent in trying to create metadata for datasets someone else created. Fortunately, the FGDC recognizes this and has a solution. First, examine Figure 5.1, which shows the general breakdown of the metadata content into nine different sections. To reduce the work in metadata creation, you can complete Sections 1 and 7 only. These two sections alone at least give future users of your dataset something to start with. Another way that the metadata workload is reduced is that numerous questions in the content standard are repeated. For the complete metadata standard, go to http://www.fgdc.gov/metadata/contstan.html.

Figure 5.1 best sums up construction the basic metadata structure. Created by the USGS's Biological Resources Division, this image has long helped users learn how the many pieces of metadata fit together.

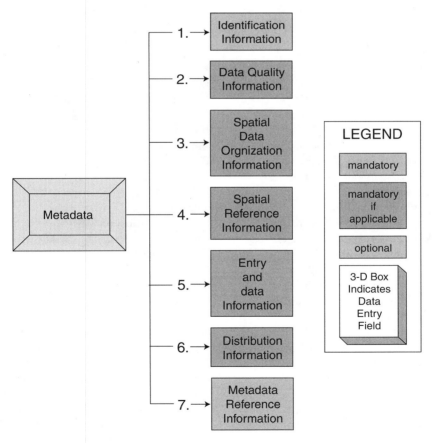

Figure 5.1. Basic metadata structure. Different shading shows which content components require data, are optional, etc.

The link to this graphic is http://www.its.nbs.gov/fgdc.metadata/version2/. Once there, click on the graphic and it will change to show similar images providing more detail on each of the major metadata components.

As an illustration of the level of information the FGDC metadata content standard encompasses, a listing of the content is shown in Table 5.2. The table was taken from Appendix B of the published FGDC Content standard.[14] Note

(text continues on page 68)

TABLE 5.2   FGDC-STD-001-1998: Appendix B
Alphabetical List of Compound Elements and Data Elements

Abscissa Resolution, 32	Azimuth Measure Point Longitude, 29
Abstract, 4	Azimuthal Angle, 28
Access Constraints, 7	Azimuthal Equidistant, 25
Access Instructions, 47	Bearing Reference Direction, 33
Address, 59	Bearing Reference Meridian, 33
Address Type, 59	Bearing Resolution, 33
Albers Conical Equal Area, 25	Bearing Units, 33
Altitude Datum Name, 35	Beginning Date, 56
Altitude Distance Units, 35	Beginning Date of Attribute Values, 40
Altitude Encoding Method, 35	Beginning Time, 57
Altitude Resolution, 35	Bounding Coordinates, 5
Altitude System Definition, 34	Browse Graphic, 8
ARC Coordinate System, 31	Browse Graphic File Description, 8
ARC System Zone Identifier, 31	Browse Graphic File Type, 8
Attribute, 38	Calendar Date, 56
Attribute Accuracy, 11	Citation, 4
Attribute Accuracy Explanation, 11	Citation Information, 53
Attribute Accuracy Report, 11	City, 59
Attribute Accuracy Value, 11	Cloud Cover, 15
Attribute Definition, 38	Codeset Domain, 40
Attribute Definition Source, 39	Codeset Name, 40
Attribute Domain Values, 39	Codeset Source, 40
Attribute Label, 38	Column Count, 18
Attribute Measurement Frequency, 41	Compatibility Information, 48
Attribute Measurement Resolution, 40	Completeness Report, 12
Attribute Units of Measure, 39	Compression Support, 47
Attribute Value Accuracy, 40	Computer Contact Information, 46
Attribute Value Accuracy Explanation, 41	Contact Address, 59
Attribute Value Accuracy Information, 40	Contact Electronic Mail Address, 60
Available Time Period, 49	Contact Facsimile Telephone, 60
	Contact Information, 58
	Contact Instructions, 60

TABLE 5.2   FGDC-STD-001-1998: Appendix B (*continued*)
Alphabetical List of Compound Elements and Data Elements

(*continues*)

**TABLE 5.2 FGDC-STD-001-1998: Appendix B (*continued*)
Alphabetical List of Compound Elements and Data Elements**

(*continues*)

TABLE 5.2   FGDC-STD-001-1998: Appendix B (*continued*)
Alphabetical List of Compound Elements and Data Elements

(continues)

TABLE 5.2   FGDC-STD-001-1998: Appendix B (*continued*)
Alphabetical List of Compound Elements and Data Elements

Source Used Citation Abbreviation, 14	Time Period of Content, 4
South Bounding Coordinate, 5	Title, 53
Space Oblique Mercator (Landsat), 26	Transfer Size, 45
Spatial Data Organization Information, 16	Transverse Mercator, 27
	Turnaround, 48
Spatial Domain, 5	Type of Source Media, 13
Spatial Reference Information, 19	Universal Polar Stereographic (UPS), 30
SPCS Zone Identifier, 31	
Standard Order Process, 43	Universal Transverse Mercator (UTM), 30
Standard Parallel, 27	
State or Province, 59	Unrepresentable Domain, 40
State Plane Coordinate System (SPCS), 31	UPS Zone Identifier, 31
	Use Constraints, 8
Status, 4	UTM Zone Number, 30
Stereographic, 27	van der Grinten, 27
Straight Vertical Longitude from Pole, 29	Vertical Coordinate System Definition, 34
Stratum, 7	Vertical Count, 18
Stratum Keyword, 7	Vertical Positional Accuracy, 13
Stratum Keyword Thesaurus, 7	Vertical Positional Accuracy Explanation, 13
Supplemental Information, 4	
Technical Prerequisites, 49	Vertical Positional Accuracy Report, 13
Temporal, 7	Vertical Positional Accuracy Value, 13
Temporal Keyword, 7	VPF Point and Vector Object Type, 18
Temporal Keyword Thesaurus, 7	
Theme, 6	VPF Terms Description, 17
Theme Keyword, 6	VPF Topology Level, 17
Theme Keyword Thesaurus, 6	VPF_Point_and_Vector_Object_ Information, 18
Time of Day, 56	
Time Period Information, 56	West Bounding Coordinate, 5

that this is only a list; in no way does it substitute for the information contained in the CSDGM. It provides a quick look at the basic content of CSDGM, but does not have the rules regarding the range of possible responses (referred to as the domain), nor does it have the detailed hierarchical organization specified in CSDGM. Before undertaking a metadata program, visit the FGDC site, contact their staff, or contact affiliated NSDI Clearinghouse locations.

Other Metadata Standards

GIS users in the United States are familiar with (or should be) the metadata standard supported by FGDC. The FGDC metadata content standard is not unique; other countries employ metadata standards to summarize the cartographic content of data exchanged in their countries. The aim of other metadata standards is the same as that defined by the FGDC. The same types of questions are generally found in the different standards, yet the content and the file structures of metadata differ by country. This makes it difficult for software to automatically read certain metadata files, and users unfamiliar with a particular country's file format will not know where in the file to look for specific data.

The International Organization for Standardization (ISO, www.iso.ch) developed an international standard for metadata. ISO Technical Committee 211 (Geographic information/Geomatics) created ISO 19115, a process for describing digital geographic datasets using a standard set of metadata elements.[15] The ISO technical committee lists 33 member countries actively participating, with 16 other countries observing.[16]

As an illustration of the international nature of the ISO standard, the Australia New Zealand Land Information Council (ANZLIC) Internet site has provided detailed summary information on the ISO standard[17]:

These elements support four major uses: discovery of data, determining data fitness for use, data access and use of data.

Discovery of data—metadata elements were selected which would enable users to locate geospatial data and also allow producers to advertise the availability of their data.

Determining data fitness for use—users can determine if a dataset meets their needs by understanding the quality and accuracy, the spatial and temporal extents and the spatial reference system used. Metadata elements were selected accordingly.

Data access—after locating and determining if a dataset will meet their needs, users may require metadata elements that describe how to access a dataset and transfer it to their site. Metadata elements were selected to provide the location of a dataset (e.g., through a URL) in addition to its size, format, price and restrictions on use.

Use of data—metadata elements were selected so that users would know how to process, apply, merge and overlay a particular dataset with others, as well as understanding the properties and limitations of the data.

The metadata data dictionary comprises the following sections, which describe metadata elements:

- Metadata
- Identification
- Data Quality
- Data Constraint
- Maintenance
- Reference System
- Spatial Representation
- Feature Catalogue
- Application Schema
- Portrayal Catalogue
- Metadata Extension
- Distribution

This data dictionary is complemented by a series of diagrams (utilising the Unified Modeling Language, UML) which graphically depict the relationships within and between the various sections. *Note*: separates material cited from ANZLIC site.

The implementation of the ISO standard is a valuable step in facilitating further exchange of geographic data. Just exchanging GIS data with the agency next door can be difficult enough. Wait until you need data from another country—data that were developed with cultural requirements, scales, methodologies, and equipment different from yours. The advantages of adopting the ISO standard are readily apparent. I like it not only for its contributions to data exchange, but also because now I do not have to provide a list of the many international metadata sites! More information on ISO 19115 can be found at the FGDC and other national geospatial sites, such as ANZLIC, in addition to the ISO site itself.

CHAPTER 6

How to Obtain GIS Data

This chapter discusses some aspects of actually getting the GIS data, that is, the actual transfer of geographic data from someone else to you. The types of data, application where it will be employed, intent of project, and so on are not issues here. However, one feature directly relating to the data, the dataset size, will obviously play a role. GIS data transfer, like moving any other digital media, ranges from the most basic "sneaker net" to experimental electronic transfer protocols such as Internet2 and wireless transfer. While I address data storage, map storage, and data transfer media, the primary means of data transfer is, and will continue to be, electronic. Electronic transfer methods are the most subject to change, and, therefore, so is this chapter (along with the Internet addresses in the appendixes). This warning aside, the chapter serves as an introduction to those of you who haven't had the pleasure of extensive GIS data downloading.

GIS Data Law 6   GIS data transfer is going to benefit directly from the advances made in digital communication. As Internet transfer software, bandwidth, data compression, and database interfaces improve, GIS data transfer will not only improve but will offer benefits to the user out of proportion to the changes in communication.

This law says that improvements in communication are more than just faster data access and transfer. The spatial components of GIS data need to interact with other data. Thus, as data transfer as a whole is improved, GIS will become much more valuable, because it can be combined with other data to present information in ways more useful to the consumer.

TRANSFER VS. STORAGE

Once you begin trading frequently in GIS data you will realize that there is more to handling GIS data than accessing and downloading it. Proper storage of the data once you receive it is very important. We can express the gathering and use of GIS as a series of steps (see Chapter 3 for more information on steps for planning GIS applications):

1. *Problem*—What problem do you have that spatial data can help solve?
2. *Application*—What is the best GIS solution?
3. *Planning*—What is the project setup and means of procuring resources?
4. *Knowledge*—What is the latest information regarding the GIS data you need?
5. *Access*—How can you locate and obtain the GIS data?
6. *Solution*—What is the proper application of the GIS data to aid in solving your problem?
7. *Storage and archive*—How do you keep and prepare the data once you are done with it?
8. *Maintenance and distribution*—What if you need future access to the data or want to share it with others?

Other chapters in the book discuss the data types, planning projects, problems and successes, and more. Those chapters support the main premise of the book, which is to put GIS users in contact with the data they need. However, the GIS data process does not stop when you have successfully used GIS to help solve another problem. The time and effort spent on a GIS project is limited (one hopes, at least), but the data gathered for the project will last forever and can always be used again, even obscure datasets. Once the data have been used it would be wise to spend a little time to keep and store the data.

Data transfer is the act of moving GIS data by whatever means from one user to another. This is as simple as handing the GIS user next to you a floppy disk or as challenging as loading 1000 DOQs for someone to download anonymously through the Internet. Data storage, on the other hand, is the process by which data are kept ready in one location for future use. The concept is nothing new; we all keep computer data of various types (word processing documents, computer game files, Internet browser bookmarks, digital photos, etc.) around for future use. I will address this practice with regard to GIS data.

Once you know what GIS data to download, where are you going to put it?

You really need to think of two immediate locations where that data will have to go: (1) on the computer, and (2) in some sort of permanent storage. This permanent storage can be further subdivided as electronic space (media) and physical space (printed maps, documents, and photographs). The computer space needed as you download data, however, is of a more immediate need: Without the space, any electronic data transfer will fail.

DATA STORAGE

We may take it as a given that you will need data storage space while working on a GIS project. But what about storage needs after the project? Many GIS practitioners specialize in addressing certain types of applications and covering specific areas. Because of this, data you obtain for one project may be useful for another. The space needed to keep the data can be placed in two groups: physical space and computer (electronic or "digital") space.

Physical space requirements for most GIS users will not be a problem. This concern primarily affects large data distribution sites operated by government agencies, libraries, universities, and companies. In these cases, storing thousands of maps and CD-ROMs is a very visible effort. Large map storage drawers and cabinets for aerial photo storage can occupy several thousand square feet of space. For smaller offices, there are a variety of display/storage racks and map tubes available. Of course, physical space is also needed for the electronic storage media. Like space for maps, plots, and photographs, the space needed for electronic storage media will not affect the vast majority of GIS users. Only the larger data storage sites need address this issue, especially if data distribution is a goal. For example, a digital orthophoto covering one 7.5-minute USGS quad requires 600 MB of storage space or approximately one CD-ROM. To cover the state of Texas requires about 4400 CD-ROMs. Add resampled data and a backup copy and the total rises to 10,000 disks! Note how a data distribution role adds to the space requirement. Frequent requests for data requires convenient access to the data. If frequent data sharing were not needed, then the data could be kept on more space-efficient media, such as tapes.

A number of choices are available for storing electronic information. Basically, the storage of electronic data (beyond the space needed on a computer to use the data) can be considered a means of *archiving* data. The term archive is used frequently in GIS in reference to data storage. Archive is defined as "a place in which public records or historical documents are preserved; also: the material preserved—often used in plural."[18] Note that archiving refers to preserving data; if the task of distributing data is added to preservation, then the data storage needs are compounded. Choices in data storage include quick-access hard disk drives, tape storage systems, digital versatile disks (DVD) and CD-ROM, and removable storage media. (While all the storage choices mentioned are all "removable" to some degree, removable storage in data transfer refers to cartridge media designed for frequent transfer between computers.)

The number of choices, the variety of solutions, and the amount of information available on these media are enormous. This book cannot begin to even outline all the information available, let alone describe all the options. The Internet is the best place to find this information and news on the rapidly changing storage options. The Internet site www.cnet.com is particularly useful. The latest information on storage devices and storage options are easily found.[19] I do want to briefly address some of the forms of storage as they pertain to GIS, however. One thing to note immediately is the manner in which different methods of storage are employed. Some storage devices, such as CD-ROMs and tapes, best serve as permanent, or archival, forms of data storage. Data written to a CD-ROM, for example, will likely not be altered and the disk serves as permanent storage for the data. Hard disk drives can serve as permanent storage, but their fast access and easy data exchange are best suited to frequently accessed data. Other data storage, such as removable storage media, can transfer data between equipment and act as temporary storage sites.

Hard Disk Drives

Changes in hard disk drives have been beneficial and welcome in the realm of GIS storage. Hard disks have long served as a very convenient, fast-access means for storing data. Until recently, modest access speeds and storage capacities limited widespread use of hard drives for large storage requirements, especially for personal computers and various Windows-based servers. Hard drives offer the most convenient form of data storage with no separate media (CD or tape, for example) required and no time-consuming sequential searches for specific data. This also adds to up to increased reliability although hard disks can fail so additional backup is recommended. The value of these devices is apparent when data have to be frequently distributed to others, particularly via the Internet. For relatively modest costs, a site can have several hundred gigabytes of GIS data on-line.*

---

*The colloquial term "on-line storage" is good for referring to fast-access storage through the Internet; a service that more GIS distribution sites are becoming familiar with.

CD-ROM and DVD

These familiar little 5.25-inch plastic platters have made a profound difference in distribution and access to large datasets. The gains in storage are readily apparent, with about 650 megabytes of data available on the CD. But I believe the real advances have been in standardization of the media and reliability. CD-ROM drives are as ubiquitous now as the then-heralded 3.5-inch floppy disks were years ago. The disks are also much more reliable. I can recall numerous problems with people trying to read data copied to floppy disks and tapes. I won't even go into the problems encountered when trying to read a single large dataset copied onto multiple floppy disks, a task that rarely seemed to work.

The key to easy distribution to a wide audience is to remove as many access variables as possible. The CD-ROM's storage capacity and ability to be accessed as simply another hard drive (a D: or E: drive) propelled data distribution into a truly viable service. CD-ROMs are the media of choice for most GIS data transfer and some archival storage. The disks can hold DOQs and satellite images—two of the largest storage fillers—and can be read by almost every PC.

Advances in CD-ROM technology, as in other media, have been ongoing. The digital versatile disk (DVD), also called the digital video disk,[20] expands the storage capacity of the basic CD platter considerably. The DVD format supports storage "capacities of 4.7 GB to 17 GB with access rates of 600 kBps to 1.3 MBps."[21] The disks are becoming more popular for holding video, whereas their employment for transferring other digital data (such as GIS) has been very gradual. The idea of using DVDs as a means for GIS data transfer has been around for several years, though their use is relatively uncommon. Time will determine their ultimate utility and we will see if these disks will eventually replace the 650-MB as the preferred storage medium for large datasets.

Removable Storage Media

There has been a favorable increase in the number of removable storage options in the last several years. Removable storage devices easily attach to your computer and can act as an additional temporary disk drive and can also be referred to as portable drives. Data are written to the removable media, usually a small cartridge, for easy transfer to another computer. While removable storage has been around for a number of years, current media hold more data, come in smaller packages, and are more reliable. Examples include "super" floppies holding up to 120 megabytes, ZIP drives with 100 and 250 megabytes of storage, and JAZ drives holding up to 2 gigabytes.

Tape Storage

The ever-present tape storage drives have also improved over time, with numerous options available to the GIS user. Several prominent tape types are used, although the digital audio tapes (DAT) are most frequently seen in general GIS data transfer and permanent backup storage. DATs are often referred to by their cartridge size, such as "8 mm" or "4 mm." The storage capacity of the tapes ranges from about 2 to 40 gigabytes. Tapes serve as good data backups and as fair data transfer media. Tape access time varies and tapes must be searched sequentially to locate specific data. The storage capacity of tapes makes them ideal for very large data distribution requirements. For larger sites with frequently accessed data, some custom solutions mix hard disk drives, tapes, and other media to combine accessibility to all data, with the most frequently requested data placed on hard disks for the fastest access and downloads.

TRANSFER: MOVING DATA

The last section discussed how to store data and how to move it around, but what about that initial transfer to your system from some remote site? I know the ordering seems backward, but you really need to know where to place data and how to move it before you download it. The introduction to storage media above is useful because many GIS datasets are not yet, and may never be, made available via electronic transfer. Some of the data you need will likely come on CD-ROM, a ZIP cartridge, or a tape.

Physical Transfer

This section briefly addresses the forms of media used to transfer large data files, such as GIS datasets. I can provide only a quick glance here. There are many, many suppliers of media and media access peripherals. Also, with the very rapid changes in computer data industry, the more I say here, the more I risk providing you with very dated material. A quick analogy: If I were discussing other electronics, such as stereos, I could still only touch on the basics (receivers vs. separate components, speaker types, etc.). I could not possibly try to list all the variations and manufacturers. And stereos are not changing anywhere near as fast as computer-related products. Check the Internet by searching under the appropriate keywords (disk drives, removable media, data transfer, DVD, etc.)

when you need the latest information in this area. Sites include CNET, ZDNET, and YAHOO.

The media used in data transfer are closely related to the fixed media used for computer storage, with the exception of the hard disk drives. Hard disk drives per se are not used to physically transfer data; removable cartridges perform this function. When you order larger GIS data sets, often the data can be provided only on some sort of physical media. CD-ROM, cartridges (ZIP and JAZ), and tapes are all used to transfer data. The CD-ROM is overwhelmingly the media of choice today. Every new computer today has a CD drive and, more importantly, virtually every computer user knows how to access and read that drive. CD reader speed has greatly increased too. While this may not seem very significant when applied to many of the common software and game programs read from CD, it makes a big difference when transferring a number of CDs all packed with data. Many programs on CD come nowhere near to filling the entire disk and frequently these files are compressed, making them smaller. An inexpensive CD drive upgrade could reduce the reading time of an entire disk by many minutes. This proved important in a data-copying exercise at work where 4000 CDs needed to be read for compression and copying. Saving 10 minutes of reading time per CD by upgrading the CD readers saved many hours of watching the data move.

This standardization has greatly reduced data transfer confusion. The ubiquity of the CD reader on all computers improved data transfer immensely. Tapes (which are still used for data access and archiving) and early cartridges held data but also came with plenty of confusion. Tape block sizes and the particular flavor of Unix used to copy the data onto the tape made it difficult for many users to easily access the data. When people cannot access data due to transfer problems like that they often look to other sources or, if the need for the data is not too great, will just return the unused tape. Even as the tape blocking and copy command concerns have been reduced by more robust software (remember the tar and cpio commands?), there are still many different sizes and types of tapes (although at least the old nine-track tapes are no longer used!). Older removable cartridges could be problematic too. Newer removable cartridges present fewer problems. New cartridge drives are designed for installation in most new computers and come with software for installation and later data transfer. The newer cartridges also hold more data, up to several gigabytes. This is more than the CD-ROM, although more widespread usage of the DVD will significantly improve on the capacity of the CD.

Another variant of the CD reader is changing data transfer: the CD writer. Having a CD writer on your computer turns you from strictly a data acceptor into a donor. GIS data are made to be shared, and having a CD writer more accessible to users everywhere greatly expands the list of possible GIS data suppliers.

Data compression is not related to media and storage devices but it has become very significant in data transfer. I can cite two areas where data compression has made strides. The first is the inclusion of simple data compression software on new computers. This software, like CD readers, is both ubiquitous and easy for most any user to apply. WINZIP is a familiar product on new PCs. Many types of files are compressed and uncompressed using WINZIP. It is particularly applicable to large text files. This has removed much of the guesswork and returns of compressed files.

The second manner in which compression has improved is in new products for compressing image files. Image file compression is not new, but having compression routines that both minimize data loss and offer ease of use are. Two compression routines with GIS applicability are MrSID (www.lizardtech.com) and ECW (www.ermapper.com). These compression packages offer compression ratios that minimize data loss up to around an 18:1 compression ratio. This means that a 154-megabyte DOQ can be compressed down to 8.5 megabytes. GIS, CAD, and image-processing software also provide routines that allow these compressed files to be read. This eliminates extra decompression steps and allows you to do all your work through the GIS software. Like having a CD writer, using this software can turn you into a GIS data distributor. The utilities allowing you to read these compressed files are free, but the software must be purchased for you to compress.

Electronic Transfer

Over the last several years, electronic transfer of GIS data have become the accepted choice. Improvements to Internet browsers and more intricate Internet pages have placed many datasets ready for instant download. Data files available for download are generally prominently displayed, often in blue text or some other distinguishing color, which you select using a computer mouse. Once a file is selected, you are asked where you would like the data file stored on your computer. Often, the Internet page showing the data provides the size of the dataset so that it can be taken into account regarding where to store the data and how long the estimated download time is.

Earlier Internet browsers also allowed downloads, although extra steps were involved, such as a "right-click" of the mouse (instead of the more common left-click) and then selecting the option to transfer the data. That was a small inconvenience though; it has been these small changes that have opened up data transfer to the millions of users who now have Internet access. The real revolution in data transfer has been through the addition of common tools that offer ease of use and are available to all computer users, not just those experienced in Unix or those with GIS software backgrounds, as was the case in the mid 1990s. Easy downloads through browsers and simple compression routines on all PCs, coupled with the CD reader, make all users participants. Once all users of technology can participate, usage explodes. Internet convenience has increased much like improvements to other technologies such as television broadcasting and communications.

FTP Transfer

File Transfer Protocol (FTP) provides standards for transferring files between computers. When data are transferred, FTP rules are followed, but the browser hides the actual computing connections, data transfer modes, and permissions. For those who have frequently transferred files prior to the widespread use of Internet browsers, computer connections and data transfers conducted through FTP are done through a more "manual" procedure. FTP interfaces can be used, which have forms for entering transfer information. The FTP connection can also be established through a DOS or Unix window. Numerous sites support data transfer via the straight FTP connections. Some sites require an account (user name and password) and some sites support downloads from any site— known as "anonymous FTP." A typical FTP session may include

- Opening a DOS or Unix window
- Typing "FTP" to enable the interface
- Typing "Open [site address]" to connect
- Typing "anonymous" under username or assigned user name if account is needed
- Typing "[e-mail address]" under password for a anonymous FTP or password for an account
- Navigating to directory where data reside, [e.g., cd /usr/public/GIS/data /hydro]
- Selecting download [e.g., "bin" for binary transfer]
- Selecting download directory on user's computer [e.g., "lcd c:/download/GIS"]
- Typing "get [filename]" to retrieve the file
- Typing "bye" to terminate the session

This was just an example of an FTP transfer session. There are other options and commands to learn about in FTP. A number of FTP guides are available through the Internet that can provide further information on FTP transfer. One example is the FTP Guide at www.scit.wlv.ac.uk/~jphb/comms/ftp.html. Other sites can be located through Internet searches. More and more file transfers are conducted through the aid of Internet browsers as discussed in the next section. Other, older computer connections, such as the BBS (bulletin board system, which is best for messages), have become obsolete for large-scale data transfer. Should you seriously want to do extensive GIS data transfers through the BBS, well, I cannot lie to you about your chances!

Internet Transfer and Speeds

Data transfer speeds through the Internet vary, depending on connections, your computer, and the Internet provider. Once you have sent information to your Internet service provider (ISP), it generally arrives quickly through the Internet backbone to the ISP of your destination (where another user then has to connect to that provider to download it). The Internet transmission speeds between ISPs along the backbone are typically at the T-3 level (45 megabits per second).[22] Your immediate connection to your ISP is the weakest link in the connectivity chain and also the one where you may have several options for improvement. The connection speed to your ISP is not the only question you have to ask when setting up a connection:

- Does your connection have the same two-way speed? That is, does it download and send information at the same speed?
- Do you expect others to download data frequently from your site? If so, you must have a fixed Internet address that does not change each time your site goes online.
- Is your Internet connection independent of existing telephone lines? This saves frequent line switching headaches and allows simultaneous use of the telephone while connected.

Table 6.1 lists some common Internet connection speeds. The connection type, speed, and download times are provided. I based the download time on our stalwart 154 megabyte CIR DOQ, a common GIS raster dataset. There are Internet sites that can tell you the speed of your own Internet connection. These

TABLE 6.1   Common Internet Connection Speeds

Connection Type	Transfer Speed	CIR DOQ Download Time
Standard 28.8 modem	28.8 kbps	712 min
ISDN (Integrated Services Digital Network)	Up to 150 kbps	136 min (maximum speed)
DSL (digital subscriber line)	384 kbps to 1.54 Mbps+ (VDSL)	13 to 53 min
Fractional T-1	128 kbps to 1.54 Kbps	13 to 160 min
T-1	1.544 Mbps	13 min
Cable modems	10 Mbps (varies with traffic)	2 min (maximum speed)
E-1, Europe	2.048 Mbps	10 min
E-3, Europe	34.368 Mbps	35 s
Fractional T-3	3 Mbps to 45 Mbps	30 s to 7 min
T-3	44.736 Mbps	30 s
OC-3	155.52 Mbps	8 s
OC-12	622.08 Mbps	2 s
OC-48	2.488 Gbps	0.5 s

sites download a test image and provide you with the connection speed. The Toast.net ISP site (www.toast.net/performance/index.html) has a good download connection speed test.

Table 6.1 is intended only as a point of reference. Speeds will continue to improve as ISPs move more into broadband communication. Broadband refers to speeds at the OC-3 (155 megabit per second speed) and higher. One of the major efforts in enhanced Internet development is the Internet2 project (www.internet2.edu) with 150 university participants and their commercial partners. The goals of Internet2 are to disseminate Internet applications to the academic, government, and commercial sectors, and to facilitate applications (medical imaging and remote participation, video, and data dissemination) not possible with current Internet technology. The Internet2 project has developed a series of gigaPoPs (gigabit Point of Presence)[23] running at speeds at OC-3 (155 megabites per second), with plans to reach the OC-196 level (9.6 gigabits per second). At 9.6 gigabits per second, the sample DOQ mentioned above could download in 0.28 seconds! For further information on advanced Internet technology see the Internet2 site or the National Science Foundation (www.nsf.gov).

CHAPTER 7

When Data Don't Exist, How Can We Make It?

The primary purpose of this book is to show you how to find and evaluate existing GIS data sources. This will not always help you, however. You will sometimes need special, or custom, data sets to solve a particular problem. Data of this nature often have very high-resolution requirements or date/time constraints. Let us pretend that you cannot locate the data you need on the Internet, get it through a government agency, or borrow it from the GIS shop down the hall. I still want this book to be of help, even for nonexistent data. By now you have the basics down for evaluating and analyzing data. If you can do that you can certainly describe the data you need and get some help in making it.

If you need to create data to complete your job, this chapter is for you. For those more interested in *selling* GIS data, however, read the following law carefully:

GIS Data Law 7 There appears to be a large difference in the relative value of GIS data (potential uses) and the absolute value of the data (what it can be sold for). It is expensive to create data, and although the data have great potential value to users, it is difficult to recoup data development costs.

If an organization spends $100,000 to develop a GIS dataset—orthophotos, for example—it likely cannot turn around and sell the data for the same amount. This is not to say that the data have lost value. Most likely, the data will be invaluable to those who created it for their own needs. Thus, the data have significant relative value for those wishing to use it. What the data do not have is the equivalent cost, or worth, equal to the costs needed to develop it. Once money has been spent to develop GIS data, the data do not seem to have an absolute value equal to or more than what it cost to make it. The simplest comparison here is noting how the value of a new automobile drops once it has been purchased and driven off the dealer's lot. Note that I said "simplest comparison"; the cost-recovery difficulty of GIS data arises from different reasons than the car's.

This apparent loss in value stems from several characteristics:

- The number of potential users for a particular GIS dataset, though increasing steadily, is relatively small with proportionally fewer customers.
- There are often similar data available through the public domain for your potential customers to access.
- Rapid changes in GIS technology make it easy to delay purchasing data now and see what other options are available.

Do not let this discussion deter you from paying for data when you need it. If the data are made correctly, they will be valuable when used in a GIS for the purchasers' needs. This generalization just throws some cautionary words to those expecting easy sales of their data.

Once you have exhausted all possible sources for existing data there are several tried and true routes for developing the data: (1) making the data yourself, (2) buying or renting it, and (3) having it custom made. The first method makes do-it-yourselfers happy, but can play havoc with administrators and the technically challenged. If you are part of an organization with considerable GIS resources, chances are you have many of the raw materials you need to do it yourself.

CREATING YOUR OWN DATA

Pros

- Reduces costs by utilizing current resources (less work done outside).
- Provides you with new experience in operations you normally don't do (or try to avoid!).
- Saves time in finding an outside source to do it for you. This is especially true if you must jump through tedious hoops like bidding projects out, government regulations, etc.
- You have control of all aspects of the project (scheduling, resources, and midstream changes).

- Sense of accomplishment (don't laugh—it's true).
- You will be an expert in many new areas when you are finished (see last con statement below).

Cons

- It can interfere with other projects and put additional pressure on everyone.
- If you are not familiar with the procedures, the time needed to learn them could be extensive.
- Additional hardware and software may have to be purchased. These things could also be used for future projects, turning these expenses into long-term benefits.
- If things go wrong, you take the blame. There is no contractor or vendor to pin it on!
- You must be an expert in all aspects of the project (production methodologies, data standards, hardware and peripherals, software versions, metadata, quality assurance); see last pro statement above.

How to Do It

One thing that you need to do no matter how you create data (make it, buy it, etc.) is follow the basic rules provided in earlier chapters on locating data. The road to developing your own data can be a rough one. The important thing is to understand why you are making it. If you are comfortable with that reason, chances are much greater that you will succeed. The reasons could include new GIS experience, a means to acquire new hardware/software otherwise unattainable, saving time, and professional pride. Build the data to fuel your GIS application first, and other benefits for creating the data will follow. Keep your priorities in order—I would not recommend building data to sell copies of it and then using it in your application. Research the general data you need, write down your requirements, and have a very clear idea what you want the data to do and what it should look like. The more you know about the data and what it can do, the easier it is to make it, buy it, or hire someone else to do it.

First, you must have a thorough understanding of the application of the data—the problem you are trying to solve—and the materials you are going to need to accomplish this. To approach this in a more organized fashion, recall the steps used in planning GIS databases presented in Chapter 3. These steps were presented in a more generic approach common to data planning; they were not discussed specifically in the context of developing your own data. Let's re-examine them in light of in-house data development. The steps are

1. Coordinate
2. Specify
3. Plan
4. Fund

5. Build
6. Distribute
7. Maintain

We can immediately discount steps 6 and 7. Your goal now is to develop data. Whether it is distributed or maintained is immaterial to this aim, although as you progress further into data development these steps will become more relevant.

*1. **Coordinate***   I would rename this step Application Knowledge. What geospatial problem will you need your new data to solve? What standards must the data adhere to? Exact specifications of the data you hope to develop must be generated. Do not make up standards unnecessarily! Standards probably exist that are at least close to whatever dataset you are making. If you are very familiar with the product you are making, you can alter the product and expand on standards sometimes. We oversaw one DOQ project where a derived DOQ product was produced that was useful to a number of users. What data format, projection, measuring units, datums, etc. are best suited for your mapping needs? Be certain, however, to leave enough room in the specifications to allow for changes not only in GIS techniques and data but also in the associated hardware and software you need.

This adds considerable additional preparation by forcing you to become an expert in several new areas. You may know every aspect related to the GIS data you need but do not know where to start in acquiring computer hardware. The opposite could just as easily be true. You do not want to be working on the "perfect" dataset for your application but be brought to a halt when your software support stops. As for changes related to GIS, there are more options than may meet the eye, and these options will change as your project progresses (unless it is a very short project). The best solution is knowledge of the field and its related data. Open as many contacts as you can, poll your colleagues, join as many listservers as you can tolerate, and visit as many applicable Internet sites as possible.

2. Specify   This is better stated as Inventory. Who will do the work and what material is needed? In-house work means that staff will have to be reassigned to complete the new tasks and additional personnel may be needed for other aspects of the work or to fill in for your current staff. The equipment and software list for the project will likely seem very substantial, but remember that careful purchasing will leave you with products useful for many subsequent projects.

3. Plan   This is the time to plan to miss no corners. By this I mean several things beyond typical project planning like timelines and milestones. Take advantage of any other related data that may help your project, even if they do not appear useful at the time. If your application requires recent orthophotos that you will have made, see what else can help. If there is an older set of orthophotos, could they be useful? Definitely! Older datasets not only provide the data (linework, images, etc.), they also can have associated ground control information and attribution that can help. Collect as much additional material as possible. Remember that GIS data often have value and uses that are not readily apparent. As with any database development planning you must have some flexibility built in for changes.

4. Fund   For data development you will need both funds and resources. The term *resources* refers not only to hardware and software, but also to personnel. Reassigning personnel may have a cost. Placing people in new role will not add to costs (raises not withstanding), but finding replacements to handle the work they used to do will. You may also need temporary help during the life of the project. Finding workers willing to work for short time periods can be difficult and it can become frustrating if they leave in the middle of the project. This doesn't apply to all the extra people needed in a project. GIS consultants are plentiful and specialize in all areas of GIS project work—planning, GIS implementation, data development, and applications. It's often harder to find the actual workers needed to make the data than it is to find a consultant.

5. Build   The building process all depends on the data you need. Scanning, digitizing, attribution, photorevising, metadata generation, and quality assurance are processes you may encounter. Since this discussion is not concerned with the particular type of data being created, I will expand briefly on major GIS data generation processes.

Scanning is done to place analog data into a digital format. Material that will become either raster (photographs, other printed images) or vector (published maps, plans, Mylar separates, and other drawings) data may be scanned. The scale, detail, and type of material scanned will dictate the type of scanner to use. When scanning, you have a lot of control over the detail captured in the scan. Use this control in maximizing the potential of the data you create. If a map must be scanned at 300 dpi, for example, it may have more uses when scanned at 600 dpi. This is an option only if your budget and data storage capacities allow it.

Scanning maps is a much faster method of entering data than manual digitizing. But scanning is sometimes only half the picture. Once map vectors are scanned, software is needed to perform the actual raster to vector conversion. Anything that is scanned will initially be a raster product. Scanned photographs and images will likely remain as raster products unless you need to extract vector features seen in the image. This is done through on-screen digitizing.

Digitizing is the manual process of transferring vector data from an analog source into a digital product. When you think of digitizing, you may envision the traditional digitizing board. A map is taped to a digitizing board with a very fine (e.g., 1/1000-inch) wire mesh embedded in the board. A digitizing "puck" is used to trace features on the map you want to record. The grid records the locations of these points. Spatial coordinates are entered for some of the digitized points and the digitizing software georeferences and projects the recorded (now digital) data. This method of digitizing is laborious and gets old very quickly with large projects. Error can be introduced through a shaky hand controlling the puck or when a map is taped loosely to the digitizing board, causing a given point on the map to slide slightly across the wire grid. Another good source for error is the map itself. You are creating data from a printed product. The printing process alone can cause error, although this is minimized when using professionally printed maps (many government map products and particularly stable base materials such as Mylar are best).

There is another type of digitizing performed when extracting vector data from raster products. On-screen, or "heads-up" digitizing is done by selecting vector features from a raster image shown on a computer screen. The computer mouse acts as the digitizing puck. You could digitize vector features from an image on a digitizing board, but images typically have no coordinates to enter. It's better to scan the image, georeference it, and then digitize from the screen. Errors in on-screen digitizing are similar to those in manual digitizing. Keeping the mouse steady and having the best scan possible minimize the error. Errors in digitizing are much more easily fixed in on-screen digitizing. You can see the last point you digitized right on the screen, whereas with manual digitizing it's often very hard to see where you selected your last point on the map. Also, with on-screen digitizing, there's no chance of taping up the map poorly.

Attribution is the process of attaching attributes to GIS data. If you have just digitized a road map, all you initially have is a set of georeferenced digital lines.

The lines have no information. GIS software is used to create a table of features that the lines need to represent. For instance, you might want to attach a road name, road type, pavement type, age, number of lanes, and maintaining organization to each line. The GIS software allows you to set up this table. These variables that are attached to the road lines are called attributes—a descriptive characteristic. The roads may not all have the same group of attributes either. You may want to have a pavement type associated with rural highways but not with urban streets. The GIS software typically has tools to help apply these tables to the lines quickly because we are not done yet. Now the actual attribute values must be placed within the table. What are the actual road names, pavement types, and numbers of lanes? These values will have to be entered unless you are creating data for another group that will complete this task.

Photorevision is the common practice of updating vector features by superimposing them over an image. This is essentially a more specialized on-screen digitizing process. Instead of creating vector data from scratch, you are modifying and correcting existing digital data.

Metadata generation is a relative newcomer (see Chapter 5), but now is an accepted, and expected, part of GIS data development. Metadata are summary data providing detailed information about the dataset you have just created. Plan for metadata before starting data development. It is much easier to create metadata through GIS software-assisted routines than to have to create them after the fact. Also, metadata are more acceptable when they are part of a data development process and not viewed as a chore to be delayed as long as possible. Creating metadata also becomes important if you wish to distribute and later maintain data (see data planning steps 6 and 7 above). Other groups using or improving on your data will need to know how you created it.

Quality Assurance is also important to future distribution and maintenance. You want to make data people can trust. Institute a program wherein data are inspected at different points in the production process. What are the data inspected against? They are compared to the data standards. This is why proper data designs and standards are so important. How else can one judge when data are being developed correctly? Research this area thoroughly prior to production. Good quality control provides a road map of how to reproduce the data. Having no or poor quality assurance may leave you with a dataset inadequate for your own needs, let alone the needs of others.

There you have it. Creating your own data is very involved. By comparison, the last two options for obtaining data are quite simple: You purchase data someone else has made or you tell someone to create exactly what you want. However, once you have gone through the process of developing your own data, you will see it is an excellent learning tool that will keep you in good stead when ordering existing data or having custom data developed.

BUYING OR RENTING DATA

Pros

- If the data you require already exist, your needs can be met very quickly. No time and effort have to be spent, by you or a vendor, to get the data ready.
- The vendor selling data wants to sell the product to as many customers as possible. This means that the seller's development costs are borne by numerous customers. The vendor does not have to depend solely on you to recover development costs. Consequently, costs are reduced.
- Some data sets being sold are continuously being updated by the vendor. New data are available when you need them.

Cons

- If you are paying to use rather than own data, your uses of the data may be restricted. For example, you may not be able to provide copies of the data to your customers, or you may be able to use the data for a limited time.
- Data that already exist may not meet all your needs. Although the vendor can usually make some changes to the data for you (projection changes, for example) changes in resolution, scale, and date may not be possible.

How to Do It

Buying existing data is a cross between locating existing data for free from numerous downloading sites and hiring a consulting company to develop the data for you from scratch. You are getting data already created, but the data are copyrighted and you must pay for the right to use it. Before discussing the types of GIS data available, let's think about some characteristics that data for sale might meet. Keep in mind the GIS law stated above: GIS is not a volume market. If you *were* to offer GIS data for sale, you would want to make it appeal to as many potential customers as possible:

- The data should be versatile to maximize potential uses by purchasers.
- The data should cover as many potential client areas of interest by being collected over a large geographic area.
- There should be some generic, straightforward data improvement processes to apply to the data to easily meet customer requirements.
- The range of dates that the data were gathered should cover as wide a range as possible.
- Costs, of course, should be as reasonable as possible.

In looking at the above criteria, it is easy to see why remote-sensing data lend themselves easily to commercial applications. Remotely sensed data have a wide range of spatial and temporal resolutions, are available for much of the earth's surface, can be easily postprocessed and improved, and are becoming more inexpensive.

Copyrighted data do exist for most data themes and types possible (e.g., GIS road and address datasets), but remotely sensed data are prevalent in the GIS data resale market. This is due to the versatility of the data. If you develop data that you wish to copyright and license or sell later, it is best if the data can serve many purposes and thus appeal to as many potential customers as possible. If copyrighted data are developed for a specific application (for example, vector data for cadastral, transportation, and facilities management applications), the potential customer base is reduced accordingly.

Because data versatility is the key, offering copyrighted data derived from satellites and aerial photography is particularly prevalent. The data include both raw (unprocessed, except for basic corrections such as registration) data and value-added data. The raw data are provided to the customer in essentially an "as is" condition with just enough processing done to make it importable into the GIS. Value-added data are created from processes that enhance the raw data and make it more readily useable for certain applications. Common examples of commercial off-the-shelf* remote-sensing data include

- Radar imagery
- Visible, infrared, and thermal imagery of various resolutions from satellites
- Aerial photographs, also including ortho rectified products

---

*The term "commercial off the shelf" (COTS) is used frequently in the information technology world. It can be applied to GIS-related software and, I believe, GIS data.

I do not want to ignore vector data. A number of vector datasets are created for commercial resale. Remember that resale GIS data need to be versatile to appeal to as wide a variety of applications as possible. Because of this, transportation and marketing data are the most commonly seen COTS vector GIS data. Transportation data have many potential users—many corporations and particularly local, regional, and state or province governments. These groups need up-to-date, accurate road data (especially with address information) and a number of companies provide these datasets. Other commercial GIS datasets include real estate data, demographics and census-related data, and aeronautical information.

For applications where it does not make sense to create and house data for site-specific studies and specialized markets, custom GIS data can be provided.

CUSTOM DATASETS

Pros

- You can order the exact GIS data you need made to your specifications.
- The vendor essentially acts as a group of experts in your employ. You determine project timetable and retain ultimate control for your project.

Cons

- Custom data creation is expensive. Your project will pay for all the vendor's expenses.
- Although the data meet your needs exactly, the data may not be useful for others with whom you want to exchange the data later.

When you employ a vendor to create a custom dataset for you, remember that the more removed the data are from standard products, the fewer potential customers the vendor will have to sell the data to others. You will bear more and more of the vendor's costs. If the vendor creates DOQs made to USGS standards (e.g., CIR (24-bit) 1-meter pixel GeoTIFF files in UTM), someone else may likely want to purchase the DOQs from the vendor, too. If you have the vendor create a 16-bit, $3\frac{2}{3}$-meter pixel DOQ in your own projection with bitmap files—well, chances are the vendor won't find anyone else looking for these DOQs!

Work with the vendor in planning your project as much as possible. Having a vendor do the work does shift the burden of being an expert in all areas of the project toward the vendor, but this does not mean ignorance is bliss. If the vendor makes a mistake and the project is late, you can blame the vendor, but you are still stuck with a late project. Think of this as a team effort and help ensure that the vendor is planning things correctly. In return, the vendor will probably spot ways to provide you with a better product than you anticipated.

CHAPTER 8

Keys to GIS Data Success

Now that you have the basic tools for locating and using GIS data, it's time to put all that knowledge to good use. But before you run out and begin to download and use data left and right, are there any other tips that may help? Yes, there are. This chapter introduces some basic traits common to getting the best of your GIS data. These ideas can help you search for data, evaluate data, or create data. One of the best overarching ideas to remember is that there is always more information about data. This goes beyond metadata.

GIS Data Law 8   The best sources for information about GIS data-related issues are people working toward the same goals you are. Locate and contact others in the field, not only experts in GIS, but experts in your field (planning, engineering, environmental management, etc.) who use GIS. The reach of GIS into everyday professions is long and becoming more common.

As with many endeavors, planning is made easier when working with others. The best single bit of advice for ensuring that you are doing the right thing with your GIS data is to talk with others in the same profession. Sometimes (probably more often that we want) the wide range of applications GIS serves, coupled with unique data and hardware/software solutions, puts us in a position where no one else is doing exactly what we are. This makes advice difficult to envision at first. From contact with GIS data developers, professionals in your field, geospatial services vendors, academia, and others, however, you will be able to combine the information they have provided and continue toward a solution. This is just a general statement; below I've provided a number of more specific thoughts that may help.

KNOW YOUR DATA NEEDS

This may sound obvious, but know exactly what you want. Write down everything you need your data to have so that you can use the data with minimal delay. Note the scale, format, projection, datum, etc. that you want the data to be in. These are all the items discussed in Chapters 3 and 4. Remember that data that meet *all* your specifications are hard to come by, but get as close as you can. The less you have to modify the data once you have them the better. Some data alterations are simple and painless (usually), such as changing the format of small vector files. Other postdownload tasks, such as having to reproject 10,000 km^2 of 1-meter DOQs, will never be forgotten!

PUT DATA IN THEIR PLACE

Have a place to put the data! So you know what data you are going to download and where to find them? The next question is, Where will you put them? Make sure you have a data directory on your computer with adequate space and a logical easy to find name (i.e., c:\GIS\download). Internet browsers will ask you where the data should go as you begin to download data. Having an empty directory will keep you from having to halt the download and find some space. You can also specify a removable storage space if you want to download to removable media, including various tapes and cartridges.

MAKE A NEW FRIEND

Get to know someone in the GIS data business. Whether you are downloading data anonymously off the Internet or stopping by a government agency, get to know the people involved. Having someone to ask for by name will really pay off when your data needs start to get a little complicated. Modern communication through e-mail, Internet browsers, and faxes makes it easy to get many data sets without ever contacting the people behind the scenes. This will work for the simpler applications but eventually you will need help. People who distribute GIS data are generally very knowledgeable about their products and want to make sure you get what you need. You will make it easier for everyone if you have taken the time to get to know one of these people. So take the time to chat beyond the usual downloading dialog. What does the data provider know about your business or agency? Who else uses the data you are looking for? What plans does the providing group have?

On a contrary note, remember that the cheerful person who has been so helpful in the past may be gone tomorrow. This is particularly true of larger agencies and companies, especially when a large, relatively anonymous group handles incoming requests on a rotating basis. Therefore, while it is great to have a particular person in mind to contact, know as much as possible about the organization in case this person moves on. If a new person at the distributing agency is inexperienced you may even be able to help him or her learn the job more quickly! This brings us to another point: Even though many GIS data are free, data exchange is really a barter system.

BARTERING FOR DATA

TANSTAAFL—This acronym, used to great effect by Robert Heinlien[24] reminds us that "there ain't no such thing as a free lunch." The most effective promotion of GIS is the exchange of data. While you are looking for data, consider what you can bring to the table. What data do your agency or company have? Would you be willing to make them available to others? Are there any services you can provide (such as software tools, papers, and links on your web site)? Anything you can do to promote data exchange will work in everyone's favor in GIS. Some ideas are

- Having software tools, advice/information, papers, or links on your web site.
- Providing data to a local, regional, or state agency for distribution to the public.
- Forming partnerships with government agencies to better exchange technology and data.

As in all business, people will remember the good along with the bad. Providing some data for your county government's web site, for example, may go a long way the next time you have trouble downloading from their site.

GOING BEYOND GIS SOFTWARE

We should assume at this point that you have GIS software sufficient to analyze the data you are about to download. The software is ready and the computer it's running on is powerful enough. You may need more than that, however. Software that can enhance your GIS can be divided into two groups: (1) data input and exchange and (2) data manipulation. Data manipulation software includes utilities for compressing and uncompressing data, reformatting and reprojecting data, and viewing tools. Data input software allows you additional methods to getting data into your GIS and often includes software you might think has nothing to do with GIS, such as word processors, spreadsheets, and databases.

LOCATING THE MOST APPROPRIATE SOURCE

As you become more familiar with searching for and locating GIS data you will notice that some data are available from multiple sources. For example, one particular dataset concerning a region in Florida may be available from a local source (municipality or county), state government, and federal agencies with business in Florida. Thus, you can go to any of these sites and retrieve the data. This redundancy is helpful should one site be unavailable, for instance. Although data may be available from more than more source, often one site is the best one for the particular dataset you are looking for. Of the multiple sites that may distribute any given dataset, one site can usually be considered the primary source, or "home" of the data. This is usually the site of the group that created and/or currently maintains the data.

If you needed transportation data for Dallas, Texas, for example, that data can likely be located though local, state, and some U.S. federal sites. There may even be international or private sources. Would it not, however, be best to get the data from the group with the most at stake in maintaining Dallas data? Look for the group that is charged with creating and maintaining that dataset. In this case, it is logical that groups in and around Dallas will have the most knowledge about any data. This would mean the City of Dallas, Dallas County, and regional organizations in the Dallas area.* This does not mean that you should always assume that the City of Dallas is the best source. The City may not be responsible for distributing its data—another governmental entity may do it or the data could be outsourced to a private company for distribution. These pe-

---

*Regional government can be hard to define. In parts of the United States, groups of counties form governmental units that have data for many regions within their boundaries. In this Dallas example, the North Central Texas Council of Governments has many Dallas-area datasets. In other countries, regional government can be handled differently. For example, Australia has municipal, state, and federal government levels; there are no county or multicounty governments to speak of.

culiarities aside, I would suggest that you at least check with the City to see who has the best copy of its data. If you do an Internet search for City of Dallas transportation data you will no doubt find multiple sources with good data, but some sites get better data (more recent, larger scales, increased numbers of features). This is very important to note if you will need to revisit Dallas sites frequently to update data.

The best rule of thumb is to go to the most local group responsible for a given area. Local and regional government and local companies are excellent sources of information. These people have the best knowledge about the area; they know where data may be out of date. They generally have the greatest stake in seeing that data are properly maintained. So while it may seem easier to go to the largest distribution nodes (in the case of Dallas these would be the Texas state government sites, Texas universities, and the USGS), make sure you are not overlooking a source closer to home. And if you do get data from an outside source, check the metadata—where did they get the data?

YOU CAN'T ALWAYS GET WHAT YOU WANT, BUT YOU CAN GET WHAT YOU NEED

Remember to maximize your return on existing data. The quantity of existing digital geographic data continues to expand. You may not always locate the exact data that you need but there are likely some datasets available that will assist in your GIS applications.

For instance, if an urban planning application in the United States called for detailed road network and land-cover data for a specific time period, it may be difficult to locate the proper data to complete the application. It may appear that a significant amount of data needs to be generated. A closer look at existing data may reveal information that at least reduces the number of data you need to generate. Finding smaller-scale datasets maintained by the state and combining that data with local road infrastructure data may solve the required road data need. If no recent land-cover maps are available, perhaps an older, smaller-scale dataset from the USGS could help. Land–cover in urban areas changes frequently, but not all land changes all the time. Older data do not necessarily mean useless data.

What about the need for data of a particular age? Finding data representing a specific time is often the biggest hurdle in using existing data sources. Two things can help here. First, data do not change as much as we think. Existing roads, boundaries, and topography are in the same location now that they were years ago. Also, related data sets can supply the necessary information. If you need a road network for New York City from 1999 but can't find one, what can be done? First, get any other road information available. A 1990 road dataset is not as up to date, but most of the roads on it will be the same in 1999 (keep in mind the scale and accuracy of substitute datasets). A more recent land-cover dataset could help too. Roads are a land cover themselves and also form boundaries between land-cover features. A recent hydrography dataset could show some road locations if the hydro data contains bridges and culverts. A boundary dataset would also contain road-influenced data, since roads often define boundaries.

START SMALL BUT THINK BIG

Know the limitations of your experience, the abilities of your GIS office, what the hardware/software can and cannot do, and the limitations of your data. When implementing GIS, plan projects that will produce tangible results but that you know you can finish. Small projects that offer a variety of GIS tasks, such as data downloading and testing, data generation, and data maintenance and distribution, allow you to gain experience without the responsibilities that come with larger projects. Larger projects have more visibility, more bosses to please, and more potential critics (they also have more potential fans, but I'm focusing on caution for now).

As subsequent GIS projects become larger, you will continue to learn new tasks but you will also be performing some functions you've done before. Try to avoid wholly experimental work (this is not always true, of course, in research projects). A huge, nationwide mapping project may look attractive, but it requires substantial funding, manpower, data throughput capacity, and time. If you have done similar projects, but on a smaller scale, these large data-mapping projects are more easily attained by scaling up a process you are already familiar with.

YOU DON'T ALWAYS GET WHAT YOU PAY FOR— SOMETIMES YOU GET MORE (OR LESS)

Think about the costs associated with getting GIS data. When you purchase data, make sure you know what you are paying for. Are you recouping someone for his or her time? Are you paying for the use of equipment? Are you paying for application of a particular technology? You need to know what costs are based on. Ask the group providing the data—whether it is a government agency, a university, or a private vendor—what the basis for their expenses is. Not only will this help you understand how costs are derived, but you will also learn something about how they conduct their business. This can be helpful in planning future projects of your own.

SHOW OFF THE PRODUCTS

This point could have easily gone into Chapter 9 (Problems) entitled "Failure to Demonstrate." Although GIS users spend lots of time working on GIS projects, the results of any GIS work must eventually be shared with others. Sometimes it can be difficult to explain GIS and its uses to the nonspatially inclined. GIS is a specialized tool with many uses, but not everyone is familiar with its utility. Let the GIS speak for itself and provide a refreshing way to look at spatial data. This could be a "pretty picture," an illustrated report, GIS graphics used in a presentation, or some other display. The people who fund GIS and the people who have to use (and benefit from) GIS data must understand what it is you are doing.

In some of our state GIS projects, a lot of effort has gone into creating graphics, handouts, Internet pages, reports, and presentations that show what GIS data are and how they are used to benefit the public. GIS benefits do not always speak for themselves; sometimes they have to be assisted. This is particularly true when trying to raise GIS-related funds or to encourage people to use data they are not familiar with. These efforts are not just for private sector sales; people in government and academia also need to promote their programs and show that GIS data are understood and used by their customers.

CHAPTER 9

Common Problems Encountered in Using GIS Data

Earlier chapters have provided basic information on GIS data: what data are out there, how to make data, some keys to success, and more. Now it is time for the bad news. What are the problems commonly found in utilizing these data? If you're guilty of these, don't worry—I am too. These things happen as one gets more experienced in using GIS. This chapter presents 10 problems that may adversely affect your GIS experience. Since I am listing some basic rules here, it is difficult to start off with another GIS law. I will include one that best summarizes the GIS problems you could encounter. The general tone of this statement may seem familiar.

GIS Data Law 9　In GIS you are mixing spatial reasoning with technology, so expect the unexpected. At any stage of GIS data operations you may encounter problems (data conversion, hardware problems, unexpected costs for software or data) that affect your ability to use GIS data. The best thing about these problems is that you are more prepared for them when they occur again.

This statement really says that GIS problems are expected but that they can always be overcome. This may sound too optimistic (sort of like referring to problems as "opportunities"), but it is true. Problems can and will come over the course of finding and using GIS data. People sometimes have trouble downloading or importing data, determining the history or pedigree of data, or getting their software to work right. I've had problems, too, but have been able to get help from other people or find ways to work around the problems. On the other hand, the future, importance, and general integration of GIS data today is as bright as ever and getting better. While problems will occur, the benefits from GIS today are far eclipsing these bumps in the road. This chapter examines some of these bumps and offers ways around them.

DATA AGE—HOW TO USE IT PROPERLY

Before knowing how best to apply your data, you must know its age. This again reinforces the importance of metadata (Chapter 5) and knowing what you want to get out of the data your are downloading (Chapter 4). Once you know the basic facts about your data, including its age, proper use of the age of data can be tough to gauge. You must be careful about using data that are too old for the application you have in mind (you don't want to find an address using a 30-year-old Los Angeles street map!). On the other hand, a map that seems out of date may be the best available information or it might be perfect for a historical application. Let's look at both of these situations separately.

EXAMPLE 1. BEST AVAILABLE INFORMATION

I can cite numerous examples where the latest GIS data for an area are not as recent as the users would like. This generally happens more often than not during any given GIS project. A typical example surfaces when you are looking for information regarding remote areas with little population and/or infrastructure that justifies the gathering of spatial information in the first place. The earth has plenty of locations where recent, or any, GIS data will be scarce. In these cases, the latest, or only, information available could be described as the best available data and then used within the constraints of its age. Remember, however, that best available data does not mean the best data. Aerial photographs from 1965 might be the latest information about a remote site, but it still may not meet your needs. Don't forget about the origin of the data either. Many GIS data have been developed from existing analog data. Note the difference in time between collecting the data and then converting it into a digital (GIS) format.

EXAMPLE 2. HISTORICAL APPLICATION

The utility of GIS is really apparent when applied to older datasets. By digitizing old analog data you can combine it with newer GIS data for change detection, historical studies, site searches, and more. One example in Texas involved digitizing the locations of shipwrecks on old navigational charts and combining these locations with newer digital navigational charts. These point data were used to see if historical references would help in narrowing down the potential search area prior to gathering in situ data.[25] Raster data are particularly well suited to historical uses because (1) it is easier to cover large remote areas through remote-sensing techniques than compiling a vector map, and (2) the analog-to-digital conversion process for historical raster data is straightforward, although you must be careful of the resulting positional accuracy.

Remember that there are many existing sources of data, with more becoming available and many being updated continually. Age is relative, however. The absolute age of the data you are considering could be taken out of context unnecessarily. There are numerous projects in the United States to create orthophotos to use as a new digital standard, serving as an update* to older paper maps. The typical orthophoto building process is a long one, however, and the photography that the orthos are based may be several years old before an orthophoto is ready for distribution. This can be a problem for catching the latest changes on the ground, but the orthophoto is updating a map that may be over 10 years old (over 20 years in some cases†). Certainly, if you have been using a 20-year-old map, one that is 3 years old can be considered an improvement!

MISUSE OF PROPRIETARY DATA

Some sources for GIS data do not sell you their data. I define sell to mean a product conveyed to you to be used as you please. Rather, some GIS sources give you a license to use their data within the confines of your organization. An agreement may, for example, list the name of a group of organizations and authorize its use within that group for its own uses without further resale, trading, or exchange of the data to other groups not listed in the agreement. This is somewhat obvious, but GIS data are commonly exchanged and shared (hence, this book!) and it is easy to think that the data on which you may have spent

---

*Notice that I say "update," not "replace." Typically, U.S. orthophotos are being created by 7.5-minute USGS quad, covering the same areas of the quad maps. This does not supercede the older paper map; these maps still have very valuable information. The orthophotos merely update the older maps. Use 'em both!

†USGS quad map users: Do not consider USGS quad maps to be obsolete! The USGS is continuously updating many of its quad maps. The USGS is also working with the future in mind by creating new digital products (e.g., orthophotos) that keep pace with the industry.

large amounts of money are yours to decide what to do with. Proprietary sources are usually associated with data created from private investment. Aerial photos, orthophotos, and satellite imagery from a private source are generally "leased" to a customer and cannot be given away or resold. This does not mean that the data are not the solution for you. Private sources can provide more custom features than those found more freely (see Chapter 7 again for a refresher) and will create data as you specify. You just need to be aware of restrictions when you order and use the data.

NOT BEING FAMILIAR WITH THE
BACKGROUND OF YOUR DATA

This refers to the history, or provenance,‡ of the data. Who created the dataset and what is it based on? This information is invaluable in deciding whether the data under consideration are adequate. If you think about it, quite a few things would be useful in ascertaining the data—projection, scale, date, creator, format, and much more. It would be good to have this information in one place, instead of having to gradually find it by experimenting with the files and looking up header records. Remember that metadata are supposed to be the source for this (see Chapter 5). Regardless of where you find data (Internet, over the counter, wherever), ask to see the metadata. Metadata can be extensive, but you do not need to look at all of it. Just check the basics for now. The usual "who, what, where, how, when" line of thinking is good here:

- *Who*—What groups or individuals created the data? Look at their contact information. You will want to call them up if you have any problems. Be familiar with citation entry sections of the metadata.
- *What*—What are the basics about the data. What scale, projection, and format are the data in? Make sure it is compatible with your GIS or at least that you can convert the data as needed.
- *Where*—This "where" doesn't refer to where the data are—you already know that. Instead, think what region the data cover. Make sure it is right for your application.

---

‡ *Webster's* refers to provenance as "origin, source." It adds a second definition: "the history of ownership of a valued object or work of art or literature." GIS data are not exactly literature, but some might consider them art—and the data are certainly valuable.

- *How*—How were the data created? Were they digitized, scanned, or updated from an earlier file?
- *When*—This is important! How old are the data? Have the original data been altered? The metadata should list all known changes to the data. If the original data have been reprojected, the metadata should note that. The extent of documenting these changes is open to some question. This can get rather confusing in that I will argue that the metadata should record all changes, even simple ones such as data reprojections. One line of thought is that if you can recreate the original data (i.e., reproject a file to its original projection), then the change does not need to be documented. I disagree with this. On the other hand, don't add to the work either. If a data set has not been altered, you don't need to cite that. Confused yet? If you need more specific information, look at the appropriate metadata section.

Look at the metadata right away—it will probably tell you more than you need to know. Of course, there must be metadata available for you to look at. Whereas it is the responsibility of the data hunter to look at the metadata carefully prior to using data, it is also the responsibility of data providers to have metadata for their users. A quick hint, if you have GIS data created by an outside group (government, private company, or consultant), always ask them to create metadata along with the data. It is much easier to create metadata in conjunction with the data.

DATA VS. SOFTWARE

This could almost be a chapter by itself. Problems of this nature usually begin after you've found what appears to be the perfect dataset. It has been downloaded or otherwise delivered and now you are ready to use it in the GIS. But it doesn't work. This happens all the time to GIS users. If it hasn't happened to you, don't worry—it will. The phrase "doesn't work" covers virtually every malady known to computing, but in GIS it usually means that the data and software are not compatible.

One common example is raster/vector data confusion. This happens when you attempt to import a raster dataset into software designed for vector data. Software used to be very limited in what it would use, but now this is changing. Software is much more versatile and will often accept more types of data. Beware that sometimes software settings need to be adjusted before certain imports, even if the software is supposed to handle that data. A problem similar to this is how the software version can affect the way data are imported. Software is generally backward compatible; that is, newer versions of the software can read data that was compatible with older versions. The opposite is generally not true, in that older software versions cannot read data designed to go with the most recent software.

The capability of software to import different data types and formats has greatly improved. This has substantially improved the usefulness of most GIS software and makes good business sense too. Having your software import competitor's files makes your product more useful to your customers, although providing capabilities to export your files in a competitor's format may be another issue.

MISAPPLYING THE DATA

Probably the most insidious form of data misapplication is using inappropriate data to reach the wrong conclusion prior to the errors being noted. This is an easy problem to fall into given the amount of data available to download. There are a number of datasets that look like they can be used for a certain application, but, in fact, are not appropriate. The only sure cures are to know your project needs and to have general experience in applying spatial data.* There are two common ways these errors may occur. The first is a simple conclusion—using the wrong data for the task. The other is in misapplying the data. One of the most prominent datasets in use worldwide is digital orthophotos (DOQs). I will use these popular data to illustrate a common data misapplication problem.

DOQs are such a popular dataset because they are becoming ubiquitous, are easy to import, have all the properties of a map, and show features on the earth's surface. The amount of data appearing on the screen may seem to provide all the detail you need. Unfortunately, even the largest-scale DOQs have their limitations. One area that GIS practitioners must be wary of entering is in applying GIS data for legal records and purposes. Legal records of a spatial record are created and updated by licensed surveyors. Surveying data are often recorded in very small units (e.g., centimeters) to a high degree of precision. Trying to replicate this detail with common GIS data can lead to mistakes. Using DOQs for

---

*By spatial applications, I'm referring to your knowledge of geographic reasoning, not so much about GIS, computers, and software. It's knowing what logically are the best data to use for a certain problem.

property boundary delineation is a good example. In the United States, the USGS standard DOQ has a pixel resolution of 1 meter. Some municipal, local, and private groups have created DOQs with half-meter, 1-foot, and even 6-inch pixels. Know the limitations of this imagery before applying it. Chapter 2 discussed GIS data types and included information on National Map Accuracy Standards (NMAS). Therefore, you know that each location provided by your GIS also comes with a degree of error. Standard 1-meter DOQs meeting NMAS still have an error not to exceed 10 meters. Even the highly detailed 1-foot DOQs (should meet NMAS at a scale of 1:2,400) have error not to exceed 4 feet. Having digital data accurate within 4 feet is beneficial, but it still does not meet the accuracy required for surveying (also remember that surveying requires specialized training and certification). Resist the temptation to draw those property boundaries over those new DOQs!

I can cite other problems stemming from reaching the wrong conclusions from your data:

- Defining shorelines from DOQs without knowing whether the water seen on the image shows the proper extent of the features you are mapping (e.g., is the reservoir you are mapping at capacity?).
- Using CAD road data as a base map for addressing and routing—you need GIS attribution and topology for this.
- Using common GIS data (scales from 1:1,200 to 1:100,000+) for engineering-level planning.
- Using DOQs for image classification without adjusting the spectral characteristics of adjacent images so that like features have similar signatures.
- Insufficient field work in identifying ground features.

The list can go on indefinitely if you are not careful.

TYPE VS. FORMAT

This confusion can be difficult to explain. A data *type* refers to a generic product created to certain widely accepted specifications. Othophotos and the USGS digital line graphs are good examples. Various software and methodologies can be used to make to make data for a given type that are similar regardless of how it was constructed. *Format* refers to changes in how data are packaged. For example, an orthophoto could be placed in BIL, TIFF, and GeoTIFF format, among others.

Different formats directly affect whether data can be imported into a GIS. For example, say your software

- Cannot read orthophotos in BIL format.
- Can import the orthophoto in GeoTIFF format with no problems.
- Can import the TIFF format, but doesn't show pixel coordinates because the TIFF format does not support the coordinates.

This is a typical experience. The best way to foresee potential problems is to know your software as well as possible. Manuals should tell what formats it can import, the formats the software keeps data in while you are manipulating it, and the formats you can export data in.

SOFTWARE VS. SOFTWARE

The simplest data applications involve importing a standard GIS data set into common GIS software. Generally, you will have fewest problems with this. As the data are manipulated, further problems will arise, especially as the data are processed through one software after the other. One example would be using a program to change or enhance a GIS data set, such as changing (or reprojecting) a file. When the data are reprojected you try to import it into the GIS where it is not recognized. Even if you contact software support, you are left with one software blaming the other for any problems you encountered. This leads to the next section.

WHERE TO GET HELP

Who do you call when you encounter problems using GIS data? Since GIS data are frequently located and downloaded somewhat anonymously over the Internet, you often never interact with the people who made the data you are about to use. Many GIS data sites also store data that they did not create. This separation between GIS data creator and user makes problem solving difficult. The simplest thing is, of course, to check with colleagues to see what they recommend. If that does not work or is not possible, where do you seek outside help? You may want to try the following sequence:

1. *Have you checked the manual?* Support personnel for all computer systems (not just GIS) will ask this first. Knowing the most about your data and the software you're trying to use will make you more knowledgeable when the people who are helping you start asking lots of questions. If you are reasonably sure the answer to your problem cannot be found in the manual, go to the next step.

2. *Call the software vendor.* It's their software, let them have the first try. Chances are they have seen this problem before. Due to the flexibility of GIS data and software, however, it is not unusual to have a problem that no one has experienced identically to date.

3. *Call the GIS data supplier.* If you used the links in this book to find and download data, you know whom to contact. Good data suppliers (whether public, private, or academic) will be willing to help you. I have always tried to point the questioner in the right direction if I don't know the answer. Good support people will always be ready to help you again if their recommendation or referral does not work—and will send you on your way with something like, "If this person can't help you, call me back and we'll try again."

4. *Post your question to a listserver.* It is always wise to subscribe to several listservers that focus on areas in which you work most closely. I subscribe to several, and although I delete numerous messages, the messages I receive sometimes provide new insight that I would have normally missed. Of course, the listservers are also available to me should I have any questions of my GIS peers. There are a number of GIS listservers that allow readers to post and answer various GIS-related questions.

COMBINING SCALES

This problem is not the same as misapplying the scale (i.e., using the right data but of the wrong scale for your application), but rather one that arises from combining GIS data with different scales. There is nothing inherently wrong with combining different scale data, but as always you must be aware of the data limitations. It has sometimes been held that combining data of different scales is forbidden. I do not agree with this and think that combining these data is often inevitable, especially given the practice of using myriad datasets downloaded from many sites.

One way to approach using data of different scales is to not combine the data, but rather assign data of differing scales to different areas of your study region. For example, land-cover data can be gathered at a variety of scales from DOQs, satellite data, and existing land-use/land-cover maps. The best solution would be to get the highest resolution data covering the entire study area, but this is often difficult to do because of costs and data availability. To create a land-cover layer for a large region, such as a state or province, 1-meter or better DOQs are expensive to create and even harder to classify. One solution would be to use large-scale data where it is needed and progressively smaller scales to cover more remote areas.

Figure 9.1 shows how scale can vary as a function of distance from a given point. In the land-cover example, the greatest scale data (for example, 1:2,400-scale DOQs) would be developed over the urban center. At the city edges the scale would change to 1:12,000-scale DOQs. Data further removed from the city would be mapped at 1:24,000, then 1:50,000, and so on. This solution provides the detail where it is needed and less detail where the surface features become sparser. A caveat to this solution is that should the city expand, the areas of higher-scale coverage will have to expand too.

A final thought on combining scales is that you must remember what really defines scale. Scale is a measure of accuracy regarding the position of data. Smaller scales do not necessarily mean fewer data. Smaller scale means that the position of the data becomes less precise. Imagine a coarse 1:1,000,000 soil map of Texas

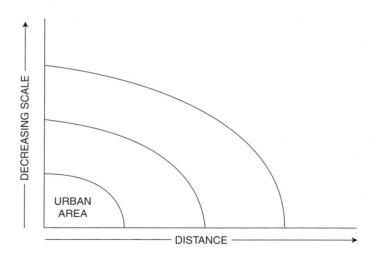

Figure 9.1. Varying scale with distance.

dividing the state into just two soil types. Now imagine overlaying a 1:24,000 road map onto the soils. From a purist's standpoint this is poor methodology, but this combination can be useful. Random points chosen within the soil map will likely be in the right soil class. Just be very careful around the borders between the two soil types.

MAKING IMMEDIATE ASSUMPTIONS

In many situations you don't want to assume too much too soon. In hunting for GIS data always be on the lookout for better data, even if you are comfortable with what you are using now. To illustrate this, the Texas Department of Transportation prints copies of the road networks in each county (254 counties for the state) in a large book of maps. While talking with a customer I learned that the roads in each county were being digitized from the maps printed in the Department of Transportation book. While progress was being made with this procedure, it was very labor-intensive! I was pleased to inform the customer that all these county roads were available in a digital (CAD) format and had been for a number of years.

This discussion on problems is not inclusive, of course. There are errors and mistakes to encounter when using and downloading GIS data. I do not want to dwell on this subject, however. Just remember that the problems and mistakes will diminish with the more GIS projects and applications you encounter.

CHAPTER 10

Future Trends

The primary aim of this chapter is to introduce future trends in spatial data distribution, but first I want to restate the basic GIS data laws found in each of the earlier chapters. I think that taken together they provide a rounded recap of the major issues that affect the idea and acts of spatial data exchange. Knowing these basics will take you a long way in maintaining effective communication with others in GIS, understanding the data you are using, knowing the geographic basics that control place and spatial relationships, and becoming a more effective GIS practitioner.

GIS Data Law 1   The reason all that existing GIS data often cannot help you is that data are usually created to solve a specific problem and are not designed to be applied to a wide range of applications.

GIS Data Law 2   If you don't know the analog data, you don't know the whole story!

GIS Data Law 3   Almost all GIS data have some value. Some data may require more manipulation but they can still make your GIS work better.

GIS Data Law 4   GIS data planning and manipulation is an ever-changing process. The amount of control GIS provides over your data is so great that you must continue to ascertain what products you want and how you want to present them throughout a project.

GIS Data Law 5   You may not be able to tell a book by its cover, but you can learn a lot about a GIS dataset from its associated metadata. Metadata are summary information (simplistically referred to as data about data) that provide a detailed description of the data. Accurate, up-to-date metadata are essential to efficient GIS data exchange.

GIS Data Law 6   GIS data transfer is going to benefit directly from the advances made in digital communication. As Internet transfer software, bandwidth, data compression, and database interfaces improve, GIS data transfer will not only improve but will offer benefits to the user out of proportion to the changes in communication.

GIS Data Law 7   There appears to be a large difference in the relative value of GIS data (potential uses) and the absolute value of the data (what it can be sold for). It is expensive to create data, and although the data have great potential value to users, it is difficult to recoup data development costs.

GIS Data Law 8   The best sources for information about GIS data-related issues are people working toward the same goals you are. Locate and contact others in the field, not only experts in GIS, but experts in your field (planning, engineering, environmental management, etc.) who use GIS. The reach of GIS into everyday professions is long and becoming more common.

GIS Data Law 9   In GIS you are mixing spatial reasoning with technology, so expect the unexpected. At any stage of GIS data operations you may encounter problems (data conversion, hardware problems, unexpected costs for software or data) that affect your ability to use GIS data. The best thing about these problems is that you are more prepared for them when they occur again.

Thus far, this book has discussed locating and downloading GIS data under current circumstances. What about the future? What will be changing in the realm of GIS data? This chapter looks at some of these changes and speculates how GIS data might be integrated into the larger information technology field. I want to limit discussion of these future trends to changes that affect data—not larger general issues, such as the Internet and information technology. In light of this, there are only two real ways in which GIS data can change with respect to accessing and using data: (1) how the data are made available, and (2) actual changes to GIS data.

  Changes in how GIS data are made available involve (1) different presentations of GIS data, (2) general technological changes to the Internet, affecting data access and delivery, (3) integration of GIS data with other information, and (4) a move toward more real-time data. The changes to the actual GIS data will include (1) new datasets, (2) updates to existing data, and (3) greater interoperability between different GIS data.

GIS DATA PRESENTATION

One of the largest generalizations that I can make is the change from a data-listing interface to an interactive map-based interface. Tied in with this are changes in how we will use regions to define our data searches. Access to most GIS datasets has historically been through various forms of lists. Using an FTP connection or an Internet browser, we are presented with a list of files to download. With FTP, we would use one of several Unix or DOS commands to download a particular file into our computer. When using Internet browsers, we could just click on a file to download it. Either way, with just a list of file names to choose from it could be difficult to determine what files to download. An Internet site can make it easier by providing appropriate directory and file names. Text files (usually named "readme" files) and metadata files are also placed in the directories for us to download and read for more detailed records of the directory contents.

List-based data access is efficient and easy to implement for the data distribution site, but it lacks details today's users need. Creative file and directory-naming conventions and readme files do make locating data to download easier, but there is still much room for improvement. The change away from the list-based selection is now being found among a number of GIS data sites on the Internet (see Chapter 6). The proper approach now is to have a user-interactive map-based interface showing regions where data are available that is accessible through the Internet. Instead of reading through a list of files, we can select the area we are interested by clicking on a map. After subsequent questions and more detailed maps our data are downloaded.

Figures 10.1 and 10.2 show the differences between list-based and interactive map-based access to data. Figure 10.1 shows a small sample list of GIS data files that could be encountered on the Internet. The files are divided by the areas they cover, in this case Harris and Fort Bend counties. The file names provide clues to the themes covered. The file extensions also show what format the data files are in. When you see the data you want, just click on the blue file name to download it. There is usually a readme file that lists facts about the files for the counties and you would download it, too, to learn about the other files in the directory. The readme file could even include full metadata on the data files.

This method is quite common and is a good way for a data site to easily store and list its holdings. Its limitations are primarily related to our having no view of the data or the area it covers. We also have to know the names of the area we wish to download. If we needed information on Houston, Texas, but didn't know the counties around the city, we could not locate the data from a

```
/data/texas/Harris_County

    harristrans.e00
    harristrans.dgn
    harrishydro.e00
    harrishydro.dgn
    boundary.e00
    readme.txt

/data/texas/Fort_Bend_County

    fortbendtrans.dgn
    boundary.e00
    fortbendsoils.e00
    readme.txt
```

Figure 10.1. File download list based on location of data within the host machine.

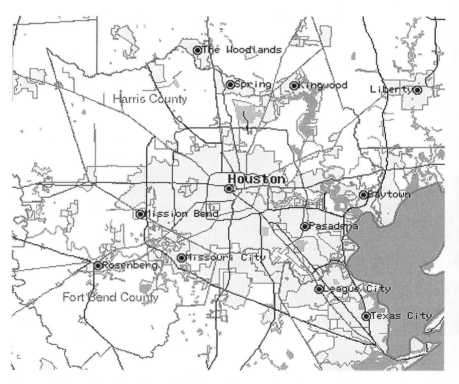

Figure 10.2. Map-based interface. Note Fort Bend and Harris counties on the map.

list based on county names. An interactive map-based interface lets us see the area we want to download first, as shown in Figure 10.2.

Look at the difference here! Instantly, we can see a map of the area we're interested in. This map-based interface allows us to select the area of interest from a map right on the screen as opposed to a list of names. We can simply click on the point where we need information or draw a box on the screen to select all data within a region. Figure 10.2 shows an area around Houston, Texas. From here, the interface may present us with questions about what data themes are needed; for example,

Select Data Type

- Transportation
- Hydrography
- Boundaries
- Land use

Subsequent manipulations through the interface would allow us to zoom in further and select a smaller area to study and then would provide a list of themes for which data are available.

Setting up distribution through a complicated interface such as this is much more work for the GIS distribution site but is very beneficial for the users. Many sites I contacted indicated that they are moving to this interface. Expect more of them. If you can easily access an on-line map to casually scan an area (for example, many sites now let you type in an address and see the road network surrounding it), why shouldn't GIS data be available in the same manner? To get GIS information truly in the hands of the public it must be presented in a way that is easily understood by virtually everyone. The information presented in the map in Figure 10.2 is obvious to everyone, but in looking at Figure 10.1, how many people know what "harristrans.e00" is? The members of the public looking for a map of Houston could easily pass "harristrans.e00" by.

CHANGES TO THE INTERNET

There is no doubt that GIS data distribution is a function of the Internet. How the Internet changes will directly affect how GIS data are distributed. By Internet, I refer to all components of the computer network—access/browser soft-

ware, data transmission speeds and hardware, networking software, mass storage, monitors, wireless transfer and more.

Data Transmission

Simply put, increased data transmission speeds will make access to large GIS
data sets much more feasible. Higher-speed connections have increased Internet
access at homes from 28.8 to over 1500 kilobits per second (1.5 megabits per
second) through digital subscriber lines (DSL)* for many U.S. telephone customers. These higher rates make raster data much more usable, but expect these
rates to continue to increase dramatically. A 154-megabyte DOQ takes 42,720
seconds (almost 12 hours) to download at 28.8 kilobits per second! Increasing
the speed by a factor of 20 through a DSL reduces that time to about 35 minutes, but that is still slow. This utility goes far beyond GIS—That 5-megabyte
upgrade for your Adobe Acrobat viewer takes 24 minutes to download with a
28.8-kilobit-per-second modem, or about 2 minutes with the DSL. Another notable competitor to the standard modem is the cable modem offered through
television cable providers. Cable modems can be faster than DSL (up to 10
megabits per second), although their speed varies with the number of users accessing the cable system at any one time.

The speeds in DSL and cable modems are in turn small alongside the speeds
available through the T-3 backbone lines used by Internet providers, agencies,
and universities. T-3 lines transfer data at 44.7 megabits per second. Faster yet
are the OC series lines used by some Internet service providers and future Internet transfer solutions through the Internet2 (www.internet2.edu) project
(Chapter 6 has more information on Internet connection speeds). The importance of transmission rate should be apparent to everyone, but what else can be
done to speed this up?

Data Compression

Data compression routines have been around for years but have been of limited
use for GIS data. This was due to competing compression formats that could not
become widely accepted by users and the fact that compression often leads to some

---

*The digital subscriber line is a marked improvement over the usual dial-up modems found on most
PCs. The DSL is much faster and operates independently of the telephone—no more switching phone
cords. The DSL helped very much in writing this book.

degree of deterioration of the data. Jack Dangermond of ESRI (Environmental Systems Research Institute) is credited with saying, "A standard ships in volume," implying that the prevailing solution or method in effect becomes industry standard even if other standards have been adopted but are seldom used.

One compression technology that is becoming increasingly accepted by GIS users is the Multiresolution Seamless Image Database (MrSID). Originally developed at Los Alamos National Laboratory, MrSID allows efficient compression of the largest raster datasets with minimal data loss or corruption. The software also allows for variable compression ratios, so data can be compressed to differing degrees. A typical 15:1 compression changes that 154-megabyte DOQ into a much more manageable 10.3-megabyte file. Complementing enhanced compression routines is free software for accessing compressed files so users do not need the compression software to view compressed files. The MrSID software is not the whole story, however. The enhanced compressed wavelet (ECW) technology developed by Earth Resource Mapping offers similar data-compression capabilities.

Mass Storage

Like data-compression programs, advances in mass storage (a.k.a. disk space) help in distributing all GIS data but are really best for raster data. Large raster files have been seemingly forbidden datasets to GIS users for years. Taking up precious disk space and confined to slow and balky tape readers for the last 30 years, large raster files are finally becoming part of the GIS data mainstream. The reasons are simple: decreased mass storage costs coupled with larger storage devices. A report from the University of California at Berkeley states that "in the past 5 years, the cost performance gap between secondary and tertiary storage has been widening. The cost per megabyte of disk drives has been falling at a factor of 2 per year, compared to 1.5 per year for tape drives and libraries."[26] In more definite terms, the cost of hard drive storage dropped from $20 per megabyte in 1988 to about $0.07 (seven cents) in 1999.[27] This trend is likely to continue, although the rate of storage cost change is difficult to predict.

What does this mean for GIS? A terabyte (1000 gigabytes) of information is not such an unusual term anymore. You can store about 6500 color infrared DOQs in a terabyte or about 2500 full *Landsat 5* scenes—a tremendous amount of data. And the cost was about $70,000 in 1999 using hard disk storage. Storing this many data solely on hard drives would be unlikely, so to shift some of that terabyte to tape storage would drive costs considerably lower. Compare this with $20,000,000 for that terabyte in 1988! Simply put, the data constantly created from previous satellites simply overwhelmed the ability of the technology of the time to store the data. This resulted in a data source that was ahead of its time relative to the computing capabilities needed to process it. Now, and even more so in the future, we can take full advantage of the data from satellites and other raster sources, such as orthophotos.

INTEGRATION OF GIS DATA WITH OTHER INFORMATION

GIS data will be increasingly "hidden" within other types of digital data. It is much easier to create an image, such as a jpeg image, immediately and provide it to the viewer rather than have the viewer directly manipulate the GIS data. The importance of this little fact is quite profound. If the power of spatial data manipulation GIS offers can be combined with the simple viewing of a jpeg image, then GIS data can be presented to everyone. Most people who need to view geographic information do not need to know anything about ESRI and Intergraph file format and projections. Data can be instantly presented in a ready to view or ready to download format. GIS data can truly be accessible (and useful!) to everyone. Users of these images, however, need to understand that the image is just a view and not true GIS data. For many people who need to show GIS output this is fine.

We need spatial data to be better integrated with the written word. As GIS users, we know that GIS is much more than "making a map." GIS provides the operations behind the creation of this seemingly simple map. For example, a GIS may create an instant map using routing and buffering techniques to show scenic routes to a hotel where you have just made reservations. You can make the reservation and receive the map/confirmation through the Internet. GIS data allow this map to be created instantly when the reservation is taken—A jpeg image is inserted into the hotel's document and it's sent in an e-mail to the future guest. The acceptance of GIS data into the larger information technology and publishing worlds will remove the misunderstood "GIS shop" facility from automatic association with geographic data.

REAL-TIME AVAILABILITY OF DATA

We have already discussed the differences between what are considered current and archival data. Sometimes the difference between the two is blurry and difficult to distinguish. The concept of "real-time" data leaves little to the imagination. These data are essentially collected while you wait. Real-time data have always been around, of course—Datasets are actively being collected right now. It's the delay between data capture and their availability to users that makes real-time data merely current.

The instant access to the Internet has now made real-time data readily accessible. Data collected by various sensors (satellites, GPS units, digital cameras, stream gauges, and traffic monitoring devices) can be sent to users directly through an Internet connection, which greatly enhances the utility of GIS data. This utility is especially apparent in monitoring and mitigating natural hazards. In its strategic plan the USGS aspires to "Ensure the continued transfer of hazards related data, risk assessments, and disaster scenarios needed by our customers before, during, and after natural disasters, and by 2005, increase the delivery of real-time hazards information by adding telemetry to 600 stream gauges . . . and installing 140 improved earthquake sensors . . . "[28]

People everywhere—not just GIS specialists—frequently rely on real-time geographic data. Good examples are the weather images from the GOES* satellites and weather radar shown on television and the Internet. To see these images, visit the National Weather Service Internet site at www.weather.gov. This site registers well over one million hits per day and has reports providing information on site usage. You can imagine how traffic at the site changes with the weather (usage during hurricane season). Another example is the real-time traffic maps now found on the Internet for some urban areas. Previously, the closest data to real time in GIS was finding the most recent satellite or aerial photography for an area. This is not the limit anymore.

In summation, the presentation of GIS data will greatly affect their impact for the better. A recent article published by the Federal Geographic Data Committee puts these changes together well:

> Vision. In the future, geodata will be ubiquitous, transparent, and "frictionless." The days of data that "just lay there" will be gone. Data will be part of transactional systems; users will receive up-to-the-minute data, along with related information such as uncertainty. Applets will come together to provide answers, not data. It will be possible to look at the local and global perspectives at the same time. This environment will move us from geodata to geowisdom.[29]

---

*Geosynchronous Operational Environmental Satellites (GOES) operated by the (U.S.) National Oceanic and Atmospheric Administration.

CHANGES TO GIS DATA

Continual improvements to the Internet are only part of the upcoming changes in GIS data. But what about changes to GIS datasets themselves? The data will change, too. Some will die out, names will change, and new datasets will be created. What kind of new GIS data can we look forward to? I can group probable dataset changes into four areas: (1) availability of larger-scale data for commonly used themes (transportation, elevation models, hydrography, facilities mapping), (2) specialized datasets not commonly used due to relatively little GIS exposure in other industries (real estate, travel planning, demographics), (3) greater accuracy and precision* due to GPS and integration with highly accurate surveying information, and (4) a greatly revitalized use of new data from satellite sources.

Larger-Scale Data

Data at larger scales are definitely making their way into today's GIS operations. Digital data first available at the 1:250,000 and 1:100,000 scales are now being made available at 1:24,000. There is no reason for scales to stop at 1:24,000, or any other scale for that matter. Improvements will always be forthcoming. The standard 1:12,000 USGS digital orthophotos have only recently (last 5 years) become a nationwide reference dataset for the United States. However, many municipalities, counties, and regional organizations have developed their own DOQs at scales of 1:6,000 and better. Vermont, for example, has based its statewide mapping program around a 1:5,000-scale DOQ base. This is all made possible by increased technology in the GIS world (GPS, digital imaging, and scanning) and in computing in general (mass storage, increased processing speeds with greatly reduced costs). The familiar transportation GIS data are moving from 1:24,000-scale linework (considered very good for the time) with positional error somewhat less than 10 meters to GPS-derived roads with accuracies of 1 meter or less. This accuracy difference is robust enough for mapping at scales of greater than 1:2,400. When introducing increased spatial accuracies and precision, with their effect on error it is wise to know the differences between these terms.[30]

---

*The terms "accuracy" and "precision" are commonly used in GIS. Let's use them right. Accuracy is a strict assessment of error. For example, a pixel in an orthophoto that is 5 meters displaced from its true position on the earth's surface is a measure of accuracy. Precision in GIS relates to the degree of detail in describing a position. A location described to the nearest foot is more precise than one described to the nearest meter. For a great description of this see "The Geographer's Craft" at the University of Colorado Geography Department Internet site. Note 5.

Specialized Datasets

Greater GIS and computing capabilities are introducing new GIS data to the public. Coupled with this is the greater exposure of GIS to more areas of the populace. This trend is easily seen in the application of GIS for human services. GIS has long been used within the world of mapping, geology, and natural resources. Its move into human services is a sign of real acceptance as a tool for solving diverse problems with a spatial component. The U.S. Census TIGER data series has had a major impact on the integration of socioeconomic data with geography. Ensuring that such a large number of data are tied to the best available spatial data has proved very difficult to do. Also, easy access to the socioeconomic information contained within has not been intuitive and has been difficult for the casual user. The Census and GIS software vendors have been working to improve this linkage. Eventually, the connections will be transparent. Geographically referenced socioeconomic data will be as cleanly presented as browsing simple html (text) Internet documents is today.

Greater Accuracy and Precision

Better technology has made it easier to increase the accuracy and precision of GIS data—no more excuses! Much of the credit for this goes to GPS. This instant location reporter is providing an easy-to-use tool to get accurate data in the field. This greatly cuts down on the guessing, map tracing, forced coordinate conversions, and other methods that people have sometimes had to employ to determine the positions within their new data. GPS is not foolproof. Some GPS receivers still require in-the-office postprocessing of the data, and acquiring the satellites needed to get a reading can be a waiting game (tree canopies, buildings, and other obstructions can block access to satellites, too). However, it is safe to say that this technology will continue to improve while costs go down. For example, GPS reception was improved in 2000 when the federal government removed distortion from the satellite signal.

The good thing about introducing this relatively easy accuracy is that, to modify a common phrase, nothing breeds accurate data like accurate data! This is saying more than that accurate data are easy to obtain; it's saying that having a core of accurate data is more likely to lead to similar data. The basic reason for this is simple: Many datasets are georeferenced to other existing datasets. The more abundant and accurate your existing data, the more accurate subsequent data will be. Data accuracy is a bit contagious—in GIS we want to work with datasets with similar scales and accuracy. Improvements in GPS positioning do not mean that GPS positions can be used for all mapping applications. The determination of property boundaries, for example, need to be addressed by licensed surveyors, regardless of the convenience of GPS. GIS can be used to store locations but these locations are only as valid as their source. The increase in GIS mapping and GPS data point collection has led to questions about the proper roles of GIS users and surveyors, promting discussion between the two groups.

Satellite Data

In this section, I'm referring to more than *Landsat* and similar satellites; I also include raster data from many sources: scanned aerial photos, airborne scanners, radar data, and elevation data. Raster data are inherently very demanding of computer storage and processing capability. For many years, the majority of GIS users have not been able to exploit the data completely. Speaking for GIS users as a whole, we are still not "there" yet. However, proper handling of these data sets is now much more feasible. Diminishing data costs coupled with home computers (not just office workstations) will allow easier access to remotely sensed data.

The satellite data itself will become more identifiable to the casual user, too. Data derived from current 1-meter resolution satellites are more appealing to the layperson and will find more applications in nonscientific uses, thus spreading their impact. Chapter 2 referred to the new *IKONOS* satellite from Space Imaging. This is the first of more to come. As more similar and improved (such as 1-meter multispectral scanners) satellites provide more frequent overpasses and lower prices, geographic data will become a staple to information hunters everywhere of all types, not just the geographically inclined.

In summation, GIS data will be going mainstream. I believe that geographic data will slowly cease to be labeled as "geographic" and will simply become data. Advances in software will greatly alleviate (if not remove, in some cases) today's GIS concerns of data incompatibility, data conversions, projections, data presentation, and file transfers. Geographic data will become less specialized, misunderstood, and mysterious and be a part of everything from HDTV viewing to e-mail. Someday the average citizen's palm computers will be dispensing geographic information with the same ease that a pager shows a phone number today. Spatial information will be not an aberration, but rather the rule. And that's when we will know geography has arrived.

Glossary

ANZLIC   Australia New Zealand Land Information Council (www.anzlic.org.au/). ANZLIC is a multijurisdiction spatial data coordination group composed of Australian state organizations and New Zealand.

Application   The particular use for which GIS data will be employed. Land-cover and transportation GIS data would be used in an urban planning application. The need for an urban planning solution dictates the application.

Archival   Older GIS data that are not thought to illustrate conditions today. There is no set time when data age and become archival data; see entry for "current."

AVHRR   Advanced Very High Resolution Radiometer. Sensor aboard NOAA satellites has five bands (visible, near infrared, mid infrared, and two thermal) that provide information on vegetation, clouds, and atmospheric conditions through analysis of reflected and emitted energy (daac.gsfc.nasa.gov/DATASET DOCS/avhrr_dataset.html).

Backward compatible   The ability of device (hardware) or software to utilize older versions of the media, software, or data that are currently used by the later versions. Examples include the ability of the latest GIS software (a trait shared with more common software, such as word processing) to load and manipulate files created with earlier versions; and the ability of the latest digital versatile disk (DVD) players to read the older CD-ROM disks.

Base map   A theme that is judged to provide essential information on common land features upon which mapping applications may be performed and from which more specialized data may be derived. Typical base mapping themes include features common to any given region, such as transportation, elevation, hydrography, land cover, and boundaries. There are few set rules on what can be a base map layer or what the scale or level of detail should be. Their determination depends on the needs of the organization developing these data for further use. The U.S. Federal Geographic Data Committee (FGDC) provides information on base map development. See *FGDC* and *them*).

Bathymetry   Isolines showing equal depth beneath a surface. Bathymetry provides the same type of information as elevation contours; here the lines usually refer to depths within a water body (lake, reservoir, ocean, or other depression). Also see *hypsography, isoline.*

Bits per second   Refers to data transfer rate. Note that the speed is measured in bits, not bytes; there are 8 bits to a byte.

CIR   Color infrared. Refers to data collected by a sensor sensitive to both visible and infrared energy. The infrared reflectance is represented by a particular color (usually red). This is more technically known as "false color infrared."

COTS   Commercial off the shelf. This term generally applies to existing software packages available for immediate sale to and use by a customer. This includes most common GIS, remote sensing, and CAD software packages. I also use the term for GIS data commercially available.

CSDGM   Content Standards for Digital Geospatial Metadata. This is the current metadata content standards document produced by the Federal Geographic Data Committee (FGDC). FGDC references the document as FGDC-STD-001-1998, *Content Standards for Digital Geospatial Metadata* (CSDGM—revised June 1998), Federal Geographic Data Committee. Washington, DC. See *FGDC* for more information.

Current   GIS data recent enough to be considered indicative of conditions today. Note that I do not use any length of time to define "currency." Current data do not have to have been gathered today; even data several years old can still provide an effective illustration of present conditions. What works depends on the data involved and the aims of your project. This same logic applies to older, archival data.

Data packaging   The practice of combining different data layers covering the same region in a manner to simplify access by the user. This is very helpful for users not accustomed to frequent GIS use and helps get more data to more people.

DCW   Digital Chart of the World. DCW is a 1:1,000,000-scale vector product providing worldwide coverage of shorelines, transportation networks, elevation, populated areas, and more, albeit with limited detail.

DEM   Digital elevation model. Grid network representing terrain by placing an elevation value at equal intervals along the grid. DEMs are created to varying scales, each with a corresponding distance between elevation points in the grid. Scales typically range from 1:24,000 to 1:1,000,000, but custom DEMs can be created for any scale or topographical application.

Desktop mapping   Software employed to create graphics designs, primarily for publications. The term is a misnomer. This is not true CAD or GIS software. It produces graphics but cannot analyze or otherwise apply the data, though often these graphics include maps.

Digitize   To convert analog data into a digital format. Traditionally performed by placing a paper map over a sensitive digitizing board and recording points taken from the paper map. This term typically refers to transferring vector data (points, lines, and areas) into digital data, but can also refer to raster digitization. Once the bane of all new GIS employees, improved scanners and conversion software have reduced (but not eliminated!) this lengthy process.

DLG   Digital line graph. Usually used to describe a digital version of one or more of the thematic layers found on the USGS quadrangle maps. Refers exclusively to vector data for layers such as transportation, elevation contours, boundaries, and hydrography. Produced at scales from 1:24,000 to 1:2,000,000.

DOI   U.S. Department of the Interior. U.S. government department charged with overseeing management of U.S. lands. A number of DOI agencies, including the U.S. Geological Survey, Bureau of Land Management, National Park Service, U.S. Fish and Wildlife Service, and Bureau of Reclamation, are actively involved in creating GIS data for federally owned lands or U.S. territory in general.

DOQ   Digital orthophoto quad. An orthophoto sized to match the area of a USGS quadrangle, typically a 3.75-minute quad (see *orthophoto*). The term DOQQ (with QQ standing for quarter-quad) is often used interchangeably with DOQ. The quarter-quad refers to the fact that the 3.75-minute DOQ (1:12,000 scale) covers one-quarter the area of the familiar 7.5-minute quad.

Download   To transfer data from one computer to another. Improvements to Internet software have made a risky function a push-button operation. Data downloaded through the Internet are typically highlighted; the user selects the data and then assigns a destination on his or her home computer.

DPI   Dots per inch. Used to describe pixel density of scanned images.

DRG   Digital Raster Graphic. A generic-sounding name referring to a specific product. DRGs are scanned versions of the 1:24,000 USGS quadrangle maps. The files are resampled to 250 dpi to balance storage and detail. Of course, other published maps that are properly scanned could serve as DRGs too. Scanned maps make excellent raster backgrounds for other, more specific data.

EDC   EROS Data Center. The Center is operated by the USGS and is the home of one of the largest sources of spatial data in the world. A vast selection of both analog and digital data is available through the EROS website (edcwww. cr.usgs.gov/eros-home.html). Digital data available include DRGs, DEMs, and DLGs. EROS also provides DOQs and satellite imagery.

ESRI   Environmental Systems Research Institute. Producers of ArcView and ArcInfo GIS software. Located in Redlands, California. The ESRI Data Hound provides links to many sites with free GIS data throughout the world. The ESRI Internet site also provides additional data resources through its ArcData page and sells data through the GIS Store (www.esri.com). ESRI has launched the Geography Network, a worldwide network of geographic data providers.

EPA   U.S. Environmental Protection Agency. Federal source for environmental impact data, pollution sources, and water resources data. See www.epa.gov.

ESIC   Earth Science Information Center. An ESIC is a USGS-supported center for the distribution of USGS products. Typically, there is one center in every state, operated by either USGS staff or a representative state entity. Regional USGS offices also operate ESICs. See mapping.usgs.gov/esic/.

FEMA   Federal Emergency Management Agency. FEMA coordinates disaster relief and litigation efforts in the United States. FEMA's best known mapping product is the 1:24,000-scale Flood Insurance Rate Map (FIRM), which shows areas at risk of potential flooding. A program is underway to update and digitize the entire set of FIRM products. See www.fema.gov.

FTP   File transfer protocol. A connection to an outside computer allowing simple functions on the host machine, including downloading files.

FGDC   Federal Geographic Data Committee. The FGDC is a federal interagency committee formed to coordinate the development of, standards for, and distribution of geographic data by most federal entities. Its better known activities include the Framework (base mapping) concept, metadata (GIS data summary information), and coordination of the National Spatial Data Infrastructure (NSDI). See www.fgdc.gov.

Georeference   Orientation of one spatial feature with respect to another based on a common coordinate system. Think of it as matching one dataset to another. For example, when importing a road vector layer over a satellite image background, the road vectors must be aligned properly over the same roads on the image. A georeferenced dataset typically has features referenced by positional coordinates.

GIS   Geographic Information System. Software designed to manage and manipulate spatial data sets.

GIS data   Information compatible with commercial, experimental, and university GIS, CAD, and remote sensing software packages. Even though the data used in GIS, CAD, and remote sensing systems differ, the geographic nature of the information allows the data to be covered by one term, thus simplifying the syntax in this book.

Heads-up digitizing   Creating vector points, lines, and areas from images on the computer screen by using a computer mouse to scribe features you want to extract. Also called on-screen digitizing.

Hypsography   Refers to elevation contours and is defined as "the scientific study of the earth's topologic configuration above sea level, especially the measurement and mapping of land elevations."[31] Digital versions of elevation contours are becoming more common. These are vector products—do not confuse them with DEMs. See *DEM, isoline, bathymetry.*

Imagery   Digital raster datasets created through electronic (nonphotographic) sensors.

Interim products   Datasets created during processing of GIS data into its final form. Typically viewed as temporary files that are used as input for further processing or are discarded. Numerous examples abound, such as temporary files generated when applying algorithms to remotely sensed imagery or when generating revised (especially photo-revised) GIS data. These datasets can be valuable either as stand-alone products or as input to other applications. The wise GIS practitioner will look to see what other uses these "temporary" data may serve.

Interoperability   Refers to the transparent integration of different datasets combined to create one dataset. For example, GIS data with different formats, projections, and datums could be combined into one dataset in the format a user specifies. Unfortunately, interoperability is still a work in progress.

Isoline   A continuous line on a map linking points of equal value. Elevation contours, for example, connect points on the ground with the same elevation.

Kbps   Thousands of bits per second. Refers to data transfer rate. With 8 bits to the byte (which is used to measure size), 1000 bits equals 125 bytes. Along the same line, Mbps represents millions of bits per second, and Gbps represents billions of bits per second.

Landsat   Earth-monitoring satellites launched between 1972 and 1999. The latest satellite is *Landsat 7* with six visible to midinfrared bands with a 30-meter ground resolution, a 60-meter-resolution thermal infrared band, and a 15-meter pancromatic band. NASA and USGS jointly manage *Landsat 7*. See landsat.gsfc.nasa.gov/.

Listserver   Computer message system whereby messages can be sent simultaneously to a large group of subscribers. Offers an excellent means to query large groups of people within one's own area of interest. GIS listservers tend to specialize, with servers devoted to imagery, popular software, standards, and more.

Metadata   Described most literally, metadata are "data about data." Metadata serve as summary information thoroughly describing a particular GIS dataset. Established standards specify the summary information to be contained within the metadata file. The best known metadata content was developed by the FGDC. Officially called the Content Standard for Digital Geospatial Metadata (CSDGM), the FGDC content is more informally referred to as "FGDC-compliant metadata." The International Organization for Standardization (ISO) is working on an International Metadata Standard that is designed to be compatible with the CSDGM so as not to make existing metadata obsolete. See www.fgdc.gov/metadata/metadata.html.

Multispectral   A sensor capable of dividing the information it records into separate groups (called *bands*). The multispectral sensor really has several different sensors, each sensitive to a certain range of reflected energy. The well-known *Landsat* multispectral scanner (MSS) had sensors that recorded reflected light energy in the green (500–600 nm), red (600–700 nm), and near infrared (700–1100 nm) bands.

NAPP   National Aerial Photography Program. U.S. federal interagency program coordinated by USGS to obtain regular photographic coverage of the United States. Photos commonly used in orthophoto production. NAPP is the current iteration of the former National High Altitude Photography (NHAP) program.

NDOP   National Digital Orthophoto Program. U.S. federal program sponsored by USGS, NRCS, and Farm Service Agency to create public domain orthophotos nationwide.

NIMA   National Imagery and Mapping Agency. U.S. Defense Department agency charged with providing geographic information to the defense and intelligence communities. Formerly called the Defense Mapping Agency (DMA) prior to restructuring in 1996.

NRCS   Natural Resources Conservation Service. Agency within the U.S. Department of Agriculture responsible for helping farmers and ranchers better manage lands by reducing erosion and conserving and protecting water. Provides extensive assistance to the proper management of private lands. Still commonly known by its former name, the Soil Conservation Service. See www.usda.nrcs.gov.

NSDI   National Spatial Data Infrastructure. NSDI describes the geographic data that are created as part of an organized, coordinated data structuring effort overseen by the FGDC. NSDI is not a group or a program, rather the geographic data are the *infrastructure*. Although the focus and beginnings of NSDI come from the federal government, NSDI is intended to support both private enterprise and all levels of government. See www.fgdc.gov/nsdi/nsdi.html.

NSGIC   National States Geographic Information Council. U.S. GIS policy and educational group primarily composed of members belonging to and representing state GIS programs. Provides forum for discussions on standards, funding, production methodologies, and data distribution. See www.nsgic.org.

Open GIS Consortium   An organization of businesses, government agencies, and universities dedicated to increased communication and compatibility among geographic information. See www.opengis.org.

Ordnance Survey   The national mapping agency for Great Britain. The Survey has analog and digital products covering a wide variety of scales. The maps are also available in series based on a particular use, such as recreation, tourism, and business travel. For more on the history of this well-known group, see www.ordsvy.gov.uk/about_us/history/.

Orthophoto   A scanned aerial photograph that is corrected to remove distortion caused by relief and attitude. The orthophoto combines the utility of a map (consistent scale, ability to derive areas, and take bearings) with the detail of a photograph.

Panchromatic   Describes a sensor that detects energy within one range, or band. Typically, the data are shown in a gray scale, or black-and-white image. The *SPOT* panchromatic sensor, for example, detects reflected energy in a range from 510 to 730 nanometers.

Raster   Data characterized by its array in a grid or cell arrangement. This includes satellite data (each *grid cell* has a specific value), elevation data (each data *point* has a certain value), and scanned images.

Raw data   Data with no or few processes done to it after it has been collected. This term is particularly applicable to remotely sensed data, such as satellite imagery and aerial photographs. Some processing may be done, such as image registration (georeferencing), but no derived products, such as classified images, ortho rectification, or thematic maps.

Remote sensing   Obtaining information about a feature without touching it. Passive remote sensing involves recording information reflected or emitted by an object (photography, satellite imaging—visible, infrared, thermal). Active remote sensing records a signal sent to and reflected by an object (radar, sonar).

Scale   Ratio of units measured on a map to the *same* units measured on the earth's surface. This is one of the most misunderstood rules in GIS and mapping in general. If the distance, for example, between two intersections on a map is one inch and the distance between the same intersections on the earth's surface (often referred to as the "real-world distance") is 100 feet, the scale is 1:1200 (one inch on the map to 1200 inches on the earth). Remember to use the same units. The measuring unit type (inches, feet, meters) does not matter, just be sure to use the same unit in measuring both the map and the earth's surface.

SPOT   *Systeme Pour l'observation de la Terre.* A French (designed by Centre National d'Etudes Spatiales, CNES) earth-observing satellite series consisting of four satellites first launched in 1984. A fifth satellite is planned for launch in 2001. The *SPOT* satellites were well known for their high-resolution 10-meter panchromatic and 20-meter multispectral scanners. See www.spotimage.fr/.

SSURGO   Recent 1:24,000-scale digital soil surveys produced by the Natural Resources Conservation Service. SSURGO coverages are created for the U.S. counties and are updated digital versions of existing surveys. The word SSURGO is derived from Soil Survey Geographic Data Base. Current SSURGO map status information is available from www.nrcs.usda.gov.

Symbology   Symbols, lines, or characters used to denote features of a certain type. Highway markers along roads on a state highway map are well-known symbology examples. Symbology also includes the components used to represent a feature. The width, color, and line type used to represent a particular type of road on a map is an example of symbology. While it is easy to accept the symbology we observe when viewing a map, designing appropriate symbology for a map you wish to print can be a laborious process.

TelNet   A connection to an outside computer that provides file access and manipulation as if the user were actually on the host machine. This is often implemented so users can run applications remotely.

Theme   The particular subject that a dataset addresses. A GIS dataset containing roads and other transportation-related features can be said to be a transportation theme. The word "theme" is used interchangeably with layer and dataset. One can have, for example, a hydrography theme, layer, or dataset. The word "dataset" typically refers to a GIS data file. The word "layer" typically refers to a GIS file in an ESRI format. Theme is a more generic term; not necessarily referring to a file or a vendor's format. For an excellent illustration of the differences of these terms, consider the digital soil surveys in production throughout the United States. Soil surveys are created by county and the total number of digital soil surveys would equal the number of counties in the country (3066). Typically, all the soil surveys together could be called a *dataset* (all 3066 files could be referred to as the soil survey dataset). A GIS soil survey file from one county may be referred to as a *layer* (the Jefferson County soil survey layer, jeff.e00). The *theme* of the data would be soils. This can be stated truthfully without regard to the number of counties or files represented.

USGS   United States Geological Survey. Primary civilian mapping agency of the United States (www.usgs.gov). An agency of the Department of Interior. The most comprehensive resource for analog and digital data, standards, printed maps, and technical information. See *EDC.*

Vector   Data composed of points, lines, and polygons (or regions).

Vectorization   Process by which vector points, lines, and area derived from a raster source, typically a scanned image.

Vendor   Term usually referring to a private company (although it can refer more generally to any outside entity) hired to perform a certain project. In GIS, these projects involve data development, photogrammetry, GIS system implementation, and custom programming.

Vertical integration   A process whereby different data layers covering the same region are properly positioned with respect to each other. For example, water features should flow under bridges, not next to them, and contours need to be properly aligned with streams to show water flowing along the bottom of a gully, not along its side.

Notes

1. Texas Geographic Information Council. "Geographic Information Framework for Texas," Texas Department of Information Resources, 1999.

2. *Webster's Revised Unabridged Dictionary*, 1998. Electronic version copyrighted © 1996, 1998 by MICRA, Inc., Plainfield, NJ. Last edit 3 February 1998.

3. Cowen, David, 1988. Taken from class lecture material, University of South Carolina.

4. U.S. Bureau of the Budget, "United States National Map Accuracy Standards," 25 October 1999; rmmcweb.cr.usgs.gov/public/nmpstds/nmas647.html, 25 October 1999.

5. National Imagery and Mapping Agency, "Digital Terrain Elevation Data (DTED®) Level 0 Frequently Asked Questions (FAQ)," 3 December 1998; 164.214.2.59/geospatial/products/DTED/faq.html 17 September 1999.

6. Jet Propulsion Laboratory, "*Landsat 6*"; msl.jpl.nasa.gov/QuickLooks/landsat6QL.html, 17 January 2000.

7. Euromap Satellitendaten-Vertriebsgesellschaft GmbH, "Indian Remote Sensing Satellite IRD-D1," 1 June 1999; www.euromap.de/doc 005.htm, 17 September 1999.

8. The Pennsylvania State University State Libraries, "About the Digital Chart of the World Data Server," 4 January 1999; www.maproom.psu.edu/dcw/dcw about. shtml, 17 January 2000.

9. Federal Geographic Data Committee, "FGDC Organization," www.fgdc. gov/fgdc/fgdc.html.

10. William Clinton, "Coordinating Geographic Data Acquisition and Access: the National Spatial Data Infrastructure," Executive Order 12906, April 1994, www.fgdc.gov/nsdi/strategy/solution.html.

11. Federal Geographic Data Committee, "GDC/ISO Metadata Standard Harmonization," updated 16 March 1999; www.fgdc.gov/metadata/whatsnew/fgdciso.html, 14 September 1999.

12. William Tolar, Federal Geographic Data Committee, personal communication, 1995.

13. Marcia McNiff, "Metadata Instruction Course," National Wetlands Research Center, 1996.

14. Federal Geographic Data Committee. FGDC-STD-001-1998. Content standard for digital geospatial metadata (revised June 1998). Washington, DC: Federal Geographic Data Committee.

15. Australia New Zealand Land Information Council, "International Metadata Standard—Revised Draft," 23 December 1999; www.anzlic.org.au/metaiso.htm, 25 January 2000.

16. International Organization for Standardization, "TC 211 Geographic information/Geomatics" www.iso.ch/meme/TC211.html, 25 January 2000.

17. "Craig Macauley, Department of Environment Heritage and Aboriginal Affairs, South Australia. International Metadata Standard, Revised Draft. Australia New Zealand Land Information Council, Canberra, www.anzlic.org.au, 1999.

18. Merriam-Webster, *WWWebster Dictionary*, 1999; http://www.m-w.com/cgi-bin/dictionary, 17 September 1999.

19. CNET, Inc., 1999; computers.cnet.com/hardware/0-1091-7-852737.html?tag=st.cn.1fd2.tlpg.1091-7-852737, 20 September 1999.

20. Internet.com, Corp., 1999; webopedia.internet.com/, 20 September 1999.

21. Internet.com, Corp., 1999; webopedia.internet.com/TERM/D/DVD.html, 20 September 1999.

22. *Internet Provider FAQ*, 1999; www.amazing.com/isp/, 20 January 2000.

23. Internet.com, 1999, *ISP Glossary*, isp.webopedia.com/, 1 February 2000.

24. Heinlein, Robert A. *The Moon Is a Harsh Mistress*. New York: Berkeley, 1966.

25. The Texas Historical Commission, "La Salle Shipwreck Project," 1999; www.thc.state.tx.us/belle/index.html, 1 September 1999.

26. University of California at Berkeley Computer Sciences Division, "The Tertiary Disk Project," now.cs.berkeley.edu/Td/">, 30 August 1999.

27. Nationwide Computer Systems, "Hard Drives," 1999; www.nationwideinc.com/harddriv.htm, 15 September 1999.

28. U.S. Geological Survey, "USGS Strategic Plan," 23 July 1999; www.usgs.gov/stratplan/, 8 August 1999.

29. U.S. Geological Survey, "FGDC Newsletter—Summer 1999," August 1999; www.fgdc.gov/publications/documents/geninfo/fgdcnl0899.html, 14 August 1999.

30. The Geographer's Craft Project, Department of Geography, University of Colorado, "Error, Accuracy, and Precision," 1 October 1997; www.utexas.edu/depts/grg/gcraft/notes/error/error.html, 5 July 2000; www.colorado.edu/geography/gcroft/contents.html.

31. *The American Heritage Dictionary of the English Language*, 3d ed. New York: Houghton Mifflin, 1996; www.dictionary.com/, 16 January 2000.

Notes on Appendixes: GIS Data Sources

The appendixes provide information and links to sources of existing GIS data around the world. Data sites are broken up by source of data (e.g., private or public, U.S. state or federal, foreign, etc.). Information on data types, web sites, and communications are provided. The sources were gathered with one primary goal: to locate where GIS data are available. Beyond the expected list of government sources for data, there are numerous private sites, too. Do not overlook private sources for data. Many companies either release free GIS data or provide links to free sites. For those who need custom GIS data developed, I have included links to GIS consultants who provide data development services.

A few words on the Internet-centric nature of the appendixes. The Internet has made finding and posting of information much easier. In fact, this instant access makes it easy to take accurate information about a given entity for granted. Usually, critical information such as phone numbers and addresses seem to lose some of their importance when so much information about a group is online. Searching, contacting, and finding information on the Internet certainly has improved from the "old days" of the Internet (1995!). The days of awkward query forms and sporadic e-mail response are gone. Web sites used to flip between being up and down continuously; also, many major organizations did not have web sites at all. Now, I can search through the Internet for any particular web site I need and be confident of finding it. Why? Because any group serious about its business (such as GIS data distribution) has a web presence. Deciding to forgo having an internet presence now is like deciding 75 years ago not to have a telephone. By internet presence, I don't necessarily mean that all groups need to have a Fortune 500 site, but they should at least be present at the level of the Internet yellow pages.

Keep in mind, however, that links to these sources are subject to change: Web site addresses change, companies merge or go out of business, even government agencies change. The contacts you may develop at these sites may change as well. That friendly person who is always willing to track down elusive data or help with downloading problems may not be there the next time you call. With these caveats in mind, the links below are the best I can find. Hope they can help you—good hunting.

WEB SITE STABILITY

One thing to note when studying lists of GIS sites on the Internet (or any sites on the Internet, for that matter) is the composition of the web site address. The address (or URL, Uniform Resource Locator) of a given Internet page provides information on who maintains the site and the general structure of the web site.

One of my goals is to provide listings of sites that we can expect to be maintained for the foreseeable future. Internet addresses do change, however, so how can we guard against this? One method is to list pages as close to the root directories of the Internet site as possible. Avoid pages that have time limitations or are dependent on the work of only one person. Examples include news stories, announcements and briefings, class material tied to a particular semester or event, references to specific discussion pages, and similar temporary information. At the other extreme, some addresses can be too vague. This is particularly true for the huge Internet holdings of large universities and U.S. federal agencies. For example,

- www.usgs.gov is stable barring any agency reorganization. We would not expect this address to change, and if it did there would be aliases and links to any new location. The USGS has enormous amounts of material on the Internet. Depending on the site design, there could be direct access to GIS data but direct access to all geographic data cannot be expected for an agency this size.
- edcwww.cr.usgs.gov/webglis is a more specific USGS Internet address. This links to the data server interface at the USGS EROS data center. A URL with a process or system name like *glis* will have to change should USGS employ a new interface and/or rename the old system. A URL with a more generic term like "data" is less likely to change.
- www.gistopics/org/education/reports/oct2000/gis_jobmarket is one detailed address! Note how information in this URL is organized logically by subject and date. This link will apparently take us to a report from released in October 2000 entitled "GIS Job Market." While this is convenient, do

not expect this report to remain under this address forever. It *could* remain there, but the chances of Internet addresses remaining permanent decrease as the subject becomes more dated. This particular report could easily be removed after several months, making this link inoperative.

I try to provide links in the appendixes that do not show signs of being changed or eliminated in the future. This cannot be prevented in all cases, of course. Company and agency names change (even large federal agencies like the former Defense Mapping Agency), organizations merge, and some sites simply are closed down. To both help track changes and provide a longer list of GIS resources, I will maintain a current on-line list of GIS resources (see the Appendixes Provided section below).

WEB SEARCHES

Searching for Internet sites has improved greatly. To direct people to any particular site, you just need to give them a few unique site-specific terms and an Internet search engine will do the rest. I recall many presentations where the audience asked the speaker to slowly repeat specific Internet addresses. No more. To find the USGS GLIS site, for example, just type in the letters G L I S, either capitals or lowercase.

ACRONYMS

Know your acronyms. Knowing GIS, governmental, and Internet acronyms will give you further insights about a site. The *edc* portion of the USGS GLIS website stands for EROS Data Center. Not everyone visiting GLIS needs to know this, of course. But it gives a clue about what is behind the web site.

PARTIAL ADDRESSES

Sometimes when attempting to link to a site you will reach a page stating that the page you are requesting cannot be found. The site itself may not be down, even though a particular page is missing. If you are looking for a USGS publication, for example,* and use the link www.usgs.gov/pubs/bulletins/2000/#107 and get an error page, try shortening the address. If you examine the address above you can see that under the primary USGS root directory a branch flows to publications. Under publications there is a link to bulletins and under that are stored bulletins from 2000. Each individual bulletin from 2000 is then grouped below this level. Try entering a shorter address than the link provided. Some ideas include the following:

- www.usgs.gov is stable, but we should be able to get closer than this.
- www.usgs.gov/pubs is more detailed. USGS will likely always support access to publications.
- www.usgs.gov/pubs/bulletins is the next step. It would be logical to group bulletins under the more general publications heading.
- www.usgs.gov/pubs/bulletins/2000 is getting trickier. Though this is getting closer to bulletin #207, these files may have been moved elsewhere, such as into an archive directory. I would still try this one first and work back if necessary.

This technique will get you closer to the section you need without necessarily having to follow the internal site routing. It is far from foolproof, however. You may arrive at a point within a web site that looks close but lacks the routing to the pages you need. Sometimes entering the site closer to the root address is better. Experience and experimentation are the best solutions.

CONTACTS

With each link I provide a phone number and an e-mail address, if available. I do this as a reminder that there is often more to an agency (or business) than we see on its Internet site. It is remarkably easy to visit a site, not find what you are seeking, and move on in search of another source. Without easy interaction with people, the site is accepted at face value. If a retail store in the mall does not have the items you want when you visit, you ask for further assistance— When will the item be restocked? Where else can I find this? Is there a substitute product? Try the same thing with GIS data on the Internet. Many sites

*The links shown in this example are examples only and not real.

show no or only partial coverage of some data layers. There could be more material that is complete and not on the site due to storage space or there could be material that is almost ready and is just waiting for someone to request it. This is particularly true with large datasets (DOQs and other imagery) and sites that house very large data collections. So, when in doubt, call or write.

APPENDIXES PROVIDED

There are six appendixes in this book:

Appendix A: U.S. Federal Sources
Appendix B: U.S. State Government Sources
Appendix C: U.S. Local and Regional Sources
Appendix D: Private Sources
Appendix E: Foreign Sources
Appendix F: Other GIS Sites of Interest

These appendixes list some of the best sources now available for GIS data. Internet searches and other investigations will provide access to the many other existing GIS data sources. There are literally thousands of sources of GIS data today. The appendixes cannot list anywhere near all the options available, so the author and John Wiley & Sons, Inc. have created an on-line GIS data resource. This Internet site will be continuously updated and supported, providing additional information for GIS data seekers. The site will have links to GIS data sources in all the categories covered by the appendixes, plus others that are felt to be of interest to the GIS profession. The site has sections for question and comments and actively seeks new listings and ideas from readers. The permanent address for the GIS Data Sources on-line site reference is www.gisdatasources.com.

U.S. Federal Sources

Name: U.S. Geological Survey (USGS)
Site: www.usgs.gov
E-mail: ask@usgs.gov
Phone: 888-ASK-USGS
Capsule: The largest and most comprehensive source for geographic data for the United States.
Notes: USGS serves as the primary civilian mapping agency for the United States. Founded in 1879, USGS has created thousands of maps, circulars, reports, bulletins, maps, aerial photographs, and digital products. Today, USGS not only funds, produces, and distributes digital products, it also has large holdings of historical maps and aerial photography. These items are valuable as archival data and are used in many time series studies. They have additional value in that many of the present-day digital data are descendents of those analog products. The USGS currently distributes ortho photography (DOQs), digital elevation models (DEMs), digital line graphs (DLGs—vector data containing base map themes), digital raster graphics (DRGs), land-cover data, and satellite data (AVHRR, *Landsat*). Data are available through several USGS sources and gateways, listed separately. USGS is very active in cooperative mapping with most federal agencies, state governments, regional government, and universities. The agency contributes funding, production capacity, expertise, and distribution and cataloging services to its partnering groups. Major USGS offices are located in Reston, Virginia (National Center), Rolla, Missouri (Mid-Continent Mapping Center), Denver, Colorado (Rocky Mountain Mapping Center), Sioux Falls, South Dakota (EROS Data Center), and Menlo Park, California (Western Mapping Center). There is enough GIS-related information on this agency site to fill this book, but I have to leave it to you to explore the huge amount of mapping information USGS possesses. Start at their homepage or visit one of the regional mapping centers.

Name: EROS Data Center (USGS-EDC)
Site: edcwww.cr.usgs.gov
E-mail: custserv@edcmail.cr.usgs.gov
Phone: 800-252-4547
Capsule: USGS digital data and remote sensing archives, including satellite data and photographs.
Notes: Most people looking for USGS digital data will eventually find the EROS Data Center. EROS is the hub for remote-sensing data maintained by USGS and other federal agencies. Its large customer service department provides digital data (DOQs, DEMs, DLGs) and can locate U.S. aerial photography and worldwide satellite data. EROS supports two major data gateways. The Global Land Information System (GLIS—edcwww.cr.usgs/webglis) offers access to common digital data series. Customers may search by data theme and scale, location, and USGS quad. Many smaller data files are free, although larger orders require a nominal fee. The EROS EOS Data Gateway (edcimswww.cr.usgs.gov/pub/imswelcome/) serves as a USGS/NASA data organization tool that provides access to *Landsat 7* and other satellite data.

Name: National Aeronautical and Space Administration (NASA)
Site: www.nasa.gov
E-mail: www.nasa.gov/hqpao/comments.html
Capsule: Many mapping and imagery resources spread throughout numerous regional sites.
Notes: Fewer groups can create as much geographical data as fast as NASA does. As everyone knows by now, NASA manages civilian space exploration, space-related research, and earth-observation data gathering. Satellites are constantly circling the earth, sending an unending stream of spatial data to ground stations. Getting those data out to all possible users, ranging from researchers to the public, is not easy. Many NASA data can be found through USGS, university-housed research centers, and related federal groups, such as the De-part-ment of Energy and NOAA. A good place to start a NASA search is at their Earth Science Enterprise (ESE) Gateway to Applications site (www.earth.nasa.gov/apps/index.html). NASA operates *Terra*, the flagship satellite of the Earth Observing System, which is designed to collect information on earth's climate by studying atmospheric processes. NASA's Gateway site provides links to Distributed Active Archive Centers (DAACs), Regional Applications Centers (RACs), Earth Science Information Partners (ESIPs), and Regional Earth Science Applications Centers (RESACs).

Name: U.S. Bureau of the Census
Site: www.census.gov
E-mail: tms@census.gov
Phone: 301-457-3030
Capsule: Home of the 10-year census data holdings and the TIGER address-based GIS data.
Notes: The U.S. Bureau of the Census is the home to an enormous supply of socioeconomic data, much of which is tied to geography. Census operates the Topologically Integrated Geographic Encoding and Referencing system (TIGER). TIGER links specific population and socially related data to geographic locations. The data consist of U.S. roads and address ranges to which various social variables may be ties. The TIGER site may be found at www.census.gov/geo/www/tiger/index.html. This page provides pricing information, data availability, and links to vendors that provide enhancements to TIGER. The Census website also has an Internet Gazetteer with connections to the TIGER Map Server (tiger.census.gov/cgi-bin/mapbrowse-tbl), one of the best on-line interactive map-generation sites. Also check out the Census Gateway (www.census.gov/geo/www/gis_gateway.html), offering connections to many GIS resources.

Name: Federal Emergency Management Agency (FEMA)
Site: www.fema.gov
E-mail: eipa@fema.gov
Phone: 202-646-4600
Capsule: Geographic information related to disaster response and mitigation. Flood plain maps.
Notes: FEMA is well known in the geographic world for its collection of Flood Insurance Rate Maps (FIRM). These maps show areas of potential flooding by plotting out the locations of 100- and 500-year flood plains. The maps are available in print from many private companies and local, regional, and state governments. The FEMA maps are being updated and digitized and are provided by county. Visit the FEMA Mapping and Analysis Center (MAC) for the latest digital mapping information (www.gismaps.fema.gov/).

Name: U.S. Fish and Wildlife Service
Site: www.fws.gov/index.html
E-mail: contact@fws.gov
Phone: 202-208-3100

Capsule: Home of the National Wetlands Inventory maps and many GIS mapping resources.

Notes: The U.S. Fish and Wildlife Service (USF&W) offers a wealth of geographic data, information on metadata and standards, data conversion tools, links, and a glossary at its GIS and Spatial Data site (www.fws.gov/data/). USF&W is also the home of the National Wetlands Inventory (NWI) map series. The NWI maps are 1:24,000-scale products available in print and several digital formats. The full-service NWI homepage offers search tools, maps to download, and map status graphics (www.nwi.fws.gov/nwi.htm).

Name: National Park Service (NPS)
Site: www.nps.gov
E-mail: infoacsm@mindspring.com
Phone: 301-493-0200, phone; 301-493-8245, fax
Capsule: Responsible for mapping national parklands.
Notes: NPS provides mapping services for parklands ranging from small parks of several thousand acres to the enormous parks in Alaska and other western states. NPS works closely with USGS and other Department of Interior agencies to create digital data layers (DLGs, DEMs, DOQs) and published maps. The large NPS site also has some unexpected touches. NPS offers an interesting little site with digital maps available of parklands in Alaska (www.nps.gov/pub_aff/lesser/lesser_frames.htm) and maps on the smaller, less visited parklands.

Name: U.S. Forest Service (USFS)
Site: www.fs.fed.us/
E-mail: www.fs.fed.us/intro/directory/fgdc.shtml [GIS contacts linking]
Phone: www.fs.fed.us/intro/directory/orgdir.shtml [agency directory]
Capsule: Responsible for mapping national forests and other U.S. Department of Agriculture properties.
Notes: Most mapping activities managed at the regional or local (state) level. Use the GIS contacts listing for GIS staff in area of interest or search for particular forest. See links to other federal mapping-related sites (www.fs.fed.us/links/maps.shtml).

Name: NASA Earth Images
Site: earth.jsc.nasa.gov
Site: images.jsc.nasa.gov
Site: eol.jsc.nasa.gov/

Capsule: NASA pages with access to spacecraft imagery.

Notes: NASA's Space Shuttle Earth Observation Program web site (earth. jsc.nasa.gov) has over 250,000 images of earth taken from space. The images are accessed through an interactive map where users select an area or a theme they want to view. The images.jsc.nasa.gov site contains an interface to NASA Space Shuttle imagery. The Office of Earth Sciences site (eol.jsc.nasa.gov) is a more general introduction to NASA's digital images of earth.

Name: National Imagery and Mapping Agency (NIMA)
Site: www.nima.mil
Capsule: Nautical, elevation, and vector data.
E-mail: chdesk@nima.mil
Phone: 314-260-5032
Notes: Formerly known as the Defense Mapping Agency, NIMA provides mapping services to the U.S. defense community. NIMA has worldwide digital vector and elevation data available for sale. See their geodata page for a complete listing (www.nima.mil/geospatial/geospatial.html).

Name: National Geodetic Survey (NGS)
Site: www.ngs.noaa.gov/
E-mail: info_center@ngs.noaa.gov
Phone: 301-713-3242
Capsule: Information on geodetic surveying, coordinate systems, GPS, and aerial photography.
Notes: NGS is a division of the National Oceanic and Atmospheric Agency (NOAA), itself a major source of coastal and bathymetric data. NGS provides information on coordinate geography, GPS, and surveying—the base elements on which GIS data depend.

Name: Bureau of Transportation Statistics (BTS)
Site: www.bts.gov/
E-mail: ntad@bts.gov
Phone: 202-366-1270
Capsule: Transportation feature-related data and information.
Notes: BTS is the data collection, analysis, and distribution arm of the U.S. Department of Transportation. Numerous transportation-related datasets, links, and standards. Home of the National Transportation Altas Data (NTAD, www.bts.gov/gis/ntatlas/natlas.html). Site has links to GIS data distribution sites

in most states. "Other resources" links reach some real-time traffic maps and general GIS information sites (see the excellent Houston real-time traffic page at traffic.tamu.edu/traffic.html).

Name: Central Intelligence Agency (CIA)
Site: www.cia.gov
Phone: 703-482-0623
Capsule: Source for unclassified printed maps worldwide.
Notes: The Central Intelligence Agency provides access to geographic information through its CIA Maps and Publications Catalog (www.cia.gov/cia/publications/mapspub/index.html). The agency has a large variety of regional and country political maps and products for sale. Some digital data are available, though most geographic data are printed. The site lists many bound publications, including the well-known annual *World Factbook*. These products are supplied to the public through the U.S. Government Publishing Office (GPO, www.access.gpo.gov/su_docs/) and the National Technical Information Service (NTIS, www.ntis.gov).

U.S. State Government Sources

This section provides information on GIS data resources in the United States available by state. Appendix A lists numerous U.S. federal sites that have GIS data nationwide, but it is important to include connections for the states as well. Why? States operate in an intermediate range between the large-scale, frequently updated, locally maintained datasets and the smaller-scale, more general federal datasets. State data follow trends initiated at the federal level, but are also tuned to residents' needs. State GIS data sites have data not available at the federal level, such as data created by state, regional, and local agencies. They also have personnel familiar with that state's GIS activities and any peculiarities with the data.

STATE SITES

States have significant leeway and initiative in creating their datasets and the means to distribute them. When looking at state data sources you will notice significant variety in how data are organized and, to a larger degree, in the organizations that support distribution. State sites are supported by universities, state agencies, interagency organizations, nonprofit groups, and state geological surveys—each state is different. Typically, one group assumes responsibility for the primary data distribution within a state (hereafter referred to as the *primary source*). This does not preclude other state groups from doing the same. When visiting the primary source you will also find links to individual agencies and groups with either additional datasets or copies of the data already available from the primary source. States each have their own GIS data-coordinating committees and personnel, support groups, and histories of GIS data development.

To keep this section from becoming too complex, I've limited my links to the primary GIS sources in each state. To further simplify this, I take advan-

151

tage in improvements in Internet searching to let my links concentrate on the basic information needed to locate the primary source. To find the state GIS data, you need the primary source name, its web address, and a telephone number if you are in too big a hurry to send e-mail. That's it. The website will have the more detailed contact information, such as staff information, fax numbers, and e-mail. I'll simply point you in the right direction, provide some comments on each state, and let you explore the state's GIS holdings from there. It's quite enough to get you started on your search; besides, you'll learn more about how each state "does" GIS.

NATIONAL SITES

But there's much more to learn about state GIS! One of the best sources is the National State Geographic Information Council* (NSGIC). Their web site is www.nsgic.org. I discuss NSGIC further in Appendix F, but it is important to think of NSGIC when researching state GIS resources. NSGIC serves as a support group for state GIS initiatives and as an information resource for state GIS practitioners. The NSGIC site lists state contacts and news, including links to many of the sites listed in this appendix. One project in particular that NSGIC is participating in is gathering detailed state profiles on all significant GIS activities for every state. At this writing, this work is in progress, but access to the completed project through NSGIC is expected by the summer of 2000.†

The Federal Geographic Data Committee (FGDC) also maintains current information on state GIS resources. FGDC sponsors the National Spatial Data Infrastructure (NSDI) Clearinghouse site, which has connections to all registered clearinghouse sites, most of which serve as state GIS data repositories (Chapter 5 has more information on FGDC, NSDI, and related initiatives, such as Framework data layers and NSDI clearinghouses). The NSDI Node Web Links page is a one-stop list showing the major GIS nodes for state and other groups. For an instant indicator of Clearinghouse server status check the following link: clearinghouse4.fgdc.gov/registry/clearinghouse sites.html, or for more general information, visit the FGDC site at www.fgdc.gov.

Another source for state GIS information is to contact a USGS Earth Sci-

---

*National States Geographic Information Council, Homepage, 22 November 1999, http://www.nsgic.org/, 14 January 2000.
†Warnecke, L., GeoManagement Associates, personal communication, 17 January 2000.

ence Information Center (ESIC). The USGS supports a group of ESICs throughout the country. An ESIC supplies analog and digital USGS for a particular region, such as a state. ESICs are operated both at regional USGS facilities and by state-related organizations. The USGS ESIC web site at <u>mapping.usgs.gov/ esic/esic_index.html</u> lists both USGS and state facilities. The ESIC concept is an older organization created for the dissemination of USGS data prior to the advent of the Internet. The current ESICs will have varying Internet capabilities. Some of the sites on the USGS list currently serve digital data (and are also listed in this appendix); other sites, however, have more limited capabilities.

Beyond NSGIC, FGDC, and ESICs (not to mention the individual states listed below), there are other sources for state data. Check Appendix D (Private Sources) to see what information some companies have made available.

VISITATION RULES

When visiting state GIS sites, it's best to keep a few questions in mind:

- What entities (local, regional, private, academic, or federal), besides the state, does this site support?
- Is this site considered the "official" site for the state? (This can be a sensitive issue depending on the local politics.)
- What other sites in the state fulfill a similar role?
- Does the site have established ties with the USGS and keep pace with federal data as it becomes available? (See the ESIC section above.)
- What types of analog (nondigital) data can be located through the site? These data include paper maps, aerial photographs, census data, published reports, atlases, statistical data, and map overlays.
- Does the site charge for the data?
- How does the site support data distribution (Internet downloads, mail, office visits, etc.)?
- What data-searching capabilities does the site support? Are data selected by theme, region, date, or some other criteria?
- What future site features are planned? If the site does not announce upcoming changes, you will have to contact the staff directly.
- What other resources does the site offer? The possibilities are vast, but if a site represents a state's GIS program it may have white papers and re-

ports, calendar of events, links to state agencies and staff, announcements for courses, training, meetings, and symposia.

Armed with these questions, it is time to begin your search for state GIS data. Remember the basics: Go to the most local GIS data source. The large federal GIS servers have lots of state data and by all means visit these federal sites for information on a particular state, but do not stop there—contact the state. Likewise, a state site may not have the best regional or local source. When considering GIS for any given piece of land, consider the entity that has the most at stake in mapping and knowing that land.

Other recommendations contained in the body of this text still hold. If you plan to visit a site frequently, make the extra effort to get to know the staff. Many datasets (such as DOQs) are so large that web site interface frequently cannot supply the data. State GIS sites may have much more data than you can find through any web site. Knowing the staff can also help when trying to track down specialized data not found in a clearinghouse, but hidden somewhere within a state agency.

LISTINGS BY STATE

Alabama
Name: Alabama Geological Survey (AGS)
Site: www.gsa.state.al.us
E-mail: www.gsa.state.al.us/gsa/ask.html [web page for submitting different types of questions]
Phone: 205-349-2852
Node: AGS is an NSDI Clearinghouse node.
Data: All available Alabama data (county, state, and federal datasets).
Notes: Data Clearinghouse page and interactive geology map server available. The site provides access to county as well as state and federal datasets. The AGS office is open to the public. AGS provides metadata for all datasets. AGS also sponsors workshops and the Alabama GIS Symposium.

Alaska
Name1: Alaska State Geo-Spatial Data Clearinghouse (ASGDC)
Site1: www.asgdc.state.ak.us
Name2: Alaska Geospatial Data Clearinghouse (AGDC)

Site2: agdc.usgs.gov
E-mail: michele@dnr.state.ak.us (ASGDC); webmaster@agdc.usgs.gov AGDC)
Phone: 907-269-8852 (ASGDC); no AGDC phone contact
Node: Both ASGDC and AGDC are NSDI Clearinghouse nodes.
Data: AGDC supports federal agency data; ASGDC supports state agency data.
Notes: Alaska supports two centralized data clearinghouses. The USGS-supported AGDC reflects the agency's strong presence in the state. ASGDC primarily supports distribution of municipal, regional, and state datasets that are located through metadata. Plans at ASGDC include obtaining and distributing more local datasets.

Arizona
Name: Arizona Land Resource Information System (ALRIS)
Site: www.land.state.az.us/alris/alrishome.html
E-mail: cblockey@lnd.state.az.us
Phone: 602-542-4060
Node: ALRIS is an NSDI node.
Data: State agency data, federal datasets (DLGs, DRGs, DEMs, and TIGER).
Notes: Metadata-driven site offers list-based search for state and federal data for Arizona. Data are supplied free via Internet, except to private entities not under contract to the state. ALRIS offers training courses to state agencies. ALRIS operates under the auspices of the Arizona Geographic Information Council (http://www.land.state.az.us/agic/agichome.html), whose site has many useful links to local and state data.

Arkansas
Name: Center for Advanced Spatial Technologies
Site: www.cast.uark.edu/local/gis/
E-mail: www.cast.uark.edu/cast/staff/email.html [web page for questions to staff]
Phone: 501-575-6159
Node: CAST is an NSDI Clearinghouse node.
Data: TIGER, USGS DOQs, DLGs, DEMs, GNIS-related data, local and state agency datasets.
Notes: The CAST site is loaded with features and links. CAST has two primary data catalogs: (1) the Arkansas Data Catalog, and (2) a National Database of Geospatial Data. It also supports the CAST Interactive Mapper, which allows maps of Arkansas to be created online. A useful document on the CAST site is

"Starting the Hunt: Guide to Mostly On-Line and Mostly Free U.S. Geospatial and Attribute Data" (www.cast.uark.edu/local/hunt/).

California
Name1: Stephen P. Teale Data Center (Teale)
Site1: www.gislab.teale.ca.gov
Name2: California Environmental Resources Evaluation System (CERES)
Site2: ceres.ca.gov/index.html
E-mail: webmaster@gislab.teale.ca.gov (Teale); ceres.ca.gov/cgi-bin/referer.pl?/comment.html [comment page]
Phone: 916-263-1767 (Teale Data Center); 916-653-5856 (California Resource Agency, CERES)
Node: The California CERES site is an NSDI node.
Data: Diverse state agency coverages, federal data (DRGs, DOQs, and DEMs).
Notes: California is different than many states in that much of the GIS coordination and data distribution is not centralized. Most progress has been made at the regional level. Teale is a private group that offers state agency and federal digital data. There are charges for data. The California Environmental Resources Evaluation System (CERES) site, operated by the California Resource Agency, offers a wide variety of socioeconomic as well as natural resource data (ceres.ca.gov/index.html).

Colorado
Name1: Rocky Mountain Geographic Information Systems Cooperative Resource Center (RMGISCRC)
Site1: www-gis.cudenver.edu/
Name2: Colorado Division of Local Government
Site2: www.dlg.oem2.state.co.us/cartog/cartog.htm
E-mail: msprain@carbon.cudenver.edu (RMGISCRC); marv.koleis@state.co.us (Colorado Local Government)
Phone: 303-626-3645 (CGCC), 303-273-1802 (Division of Local Government)
Node: No.
Data: TIGER, local boundaries and state districts, shaded relief maps.
Notes: The Colorado Geographic Information Directory is not a data distribution site, but rather offers connections to Colorado GIS groups. The Colorado Division of Local Government supplies local data (boundaries, populations, and districts). There is not a general statewide distribution node for Colorado, although this is offset by the fact that Denver, Colorado, is home to the USGS' Rocky Mountain Mapping Center, which supports an ESIC.

Connecticut
Name: Map and Geographic Information Center (MAGIC)
Site: www.uconncgia.uconn.edu/
E-mail: tyler@neca.com
Phone: 860-486-4589
Node: MAGIC is an NSDI Clearinghouse node.
Data: Wide variety of local, regional, state, multistate (New England), and federal datasets.
Notes: MAGIC operates a very comprehensive data server providing detailed local, state, and federal data. Metadata are included for all datasets. Both data images and Arc/Info files are available for downloading. Whereas MAGIC is the state's primary data provider, UCCGIA is a GIS center with an emphasis on the research applications of GIS and operates in conjunction with MAGIC.

Delaware
Name: Delaware Spatial Data Clearinghouse
Site: gis.smith.udel.edu/fgdc2/clearinghouse/
E-mail: gis.rdms.udel.edu/fgdc2/clearinghouse/comments.html [comments web page]
Phone: 302-739-3090
Node: The site is an NSDI Clearinghouse node.
Data: Statewide framework data layers
Notes: The Delaware site offers data, reports, information on state GIS activities, and links to other state and federal sites. The statewide coverages are presented as individual framework layers. Like some other states, Delaware links its data searches through the NSDI Clearinghouse search site.

Florida
Name: Florida Geographic Information Board (GIB)
Site: als.dms.state.fl.us/~fdd/
E-mail: webmaster@als.dms.state.fl.us
Phone: 850-488-7986
Node: The Florida GIB is an NSDI Clearinghouse node.
Data: Very extensive holdings of state and the full USGS suite of digital data.
Notes: The Florida Data Directory (FDD) is operated as a service of the Florida Geographic Information Board (GIB). The site is a metadata-based data server that offers general digital datasets and offers a separate catalog of aerial photography. Data may be located by theme, region, date, and type. Data queries pro-

vide an image of the region covered by the data in additional to downloading links. FDD is a full-service data library and is backed by extensive reference materials and documentation.

Georgia
Name: Georgia GIS Data Clearinghouse
Site: www.gis.state.ga.us/
E-mail: gischouse@state.ga.us
Phone: 404-894-0502
Node: Site is an NSDI Clearinghouse node.
Data: County and statewide datasets, full USGS suite of digital data.
Notes: The Georgia Clearinghouse has a very comprehensive catalog of statewide and county-based datasets. Users may select data by theme, region, USGS quad, and keyword. The site has other features, including data location tools, FAQ, links, and lots of general information on data and how to access it. Data available via downloading; larger datasets must be sent on CD. Additional GIS data nodes are found at Georgia Tech and the University of Georgia.

Hawaii
Name: Hawaii Statewide Planning and Geographic Information System
Site: www.hawaii.gov/dbedt/gis/
E-mail: dkim@dbedt.hawaii.gov
Phone: 808-586-2423 (agency switchboard)
Node: This Hawaii site is not an NSDI Clearinghouse node.
Data: Primarily municipal and state agency data, some USGS digital datasets (DEMs, DLGs).
Notes: The Hawaii Department of Business Economic Development and Tourism maintains this site, which has links to other Hawaii GIS resources, including state GIS organizations. Data are listed in several categories (base map, boundaries, natural resources, and hazards) and are available for download.

Idaho
Name: Idaho Geological Survey (IGS)
Site: www.uidaho.edu/igs/igs.html
E-mail: igs@uidaho.edu
Phone: 208-385-4002

Node: Neither the IGS nor the Idaho Geospatial Data Center (below) is an NSDI Clearinghouse node.
Data: DEM and DLG data for all quads.
Notes: The Idaho Geological Survey concentrates on digital geologic and mining data. The Idaho Geospatial Data Center housed at the University of Idaho (geolibrary.uidaho.edu/) has more general GIS coverages, such as DLGs and DEMs, DRGs, and TIGER data.

Illinois
Name: Illinois Natural Resources Geospatial Information Clearinghouse (INR)
Site: www.isgs.uiuc.edu/nsdihome/ISGSindex.html
E-mail: denhart@isgs.uiuc.edu
Phone: 217-333-4747
Node: INR is an NSDI Clearinghouse node.
Data: State agency (statewide), local data. Most offered at 1:24,000 scale.
Notes: INR offers access to all levels of data covering Illinois—local, state, and federal. INR has also teamed up with the Illinois State Museum to distribute data. Other data agencies (DNR, DOT, State Nuclear Safety) also maintain data distribution sites. Data primarily distributed via Internet and mail (CD). INR is continuously making more data available at 1:24,000. One of the strengths of INR is the access to the sometimes hard to find local data.

Indiana
Name: Center for Advanced Applications in Geographic Information Systems (CAAGIS)
Site: danpatch.ecn.purdue.edu/~caagis/
E-mail: caagis@ecn.purdue.edu
Phone: 765-494-5954
Node: No.
Data: General themes provided for county, watershed, and state regions. Includes soil surveys.
Notes: Purdue University server providing access via an Internet map server to county, regional (watersheds), and state coverages. County data include highways, roads, rails, transportation lines, streams/rivers, general depth-to-watertable, basement geology, hydrologic setting, effective clay thickness, wetlands, and STATSGO (regional) and SSURGO (1:24,000) soil data. State data

cover similar themes and add glacial geology, land use, digital elevation map, hydrologic unit areas, and Indiana hydrology.

Iowa
Name: Iowa GIS Clearinghouse
Site: www.gis.state.ia.us/
E-mail: gis@its.state.ia.us
Phone: 515-281-6137
Node: No.
Data: Links to Iowa federal, regional, county, local and statewide datasets.
Notes: The Iowa Clearinghouse is supported by the Iowa Geographic Information Council (IGIC). The site has information on state workshops, conferences, and other GIS-related activities. It also has links to IGIC members and staff. The site offers an interactive map server, allowing the user to select from a number of themes and build statewide maps or maps for selected counties.

Kansas
Name: Data Access and Support Center (DASC)
Site: gisdasc.kgs.ukans.edu/dasc.html
E-mail: dasc@kgs.ukans.edu
Phone: 785-864-3965
Node: DASC is an NSDI Clearinghouse node.
Data: USGS digital data, state agency data.
Notes: DASC supplies data via the Internet and also through custom services. This site serves as the primary source of Kansas data, along with smaller sites operated by state agencies. No local data are available, although that service is planned. There are generally no charges for data, though this policy varies with the organization requesting data. DASC offers custom map creation and data conversion and tiling. Plans include seamless data, Internet-based mapping. The site is operated by the Kansas Geological Survey.

Kentucky
Name: Kentucky Office of Geographic Information Systems (OGI)
Site: ogis.state.ky.us/
E-mail: ogis.webmaster@mail.state.ky.us
Phone: 502-573-1450
Node: OGI is not an NSDI Clearinghouse.

Data: USGS DOQs, DRGs, DEMs, and 1:24,000 DLG layers. NRCS SSURGO soils.

Notes: An informative and detailed site with numerous standard USGS digital products available to download. No charge for data unless large orders need to be sent on CD-ROM. Detailed information on state GIS standards. The OGI site also has many links to the numerous local, state, and federal offices in Kentucky with GIS data and other resources. One standout product is mosaiced, seamless, compressed DOQs.

Louisiana
Name: Louisiana Geographic Information Center
Site: lagic.lsu.edu/
E-mail: None specified
Phone: 225-388-3479
Node: LAGIC is an NSDI node.
Data: University, state agency, and federal (varying themes, 1:24,000 and 100,000 coverages).
Notes: LAGIC is housed at Louisiana State University (LSU) and has links to university data and research center. LAGIC supplies data through links to outside sources where data are stored. Quick searches by agency or keyword. A wide variety of data are available with varying themes and scales. In addition to LAGIC, LSU and the state's Department of Environmental Quality offer significant GIS data.

Maine
Name: Maine Office of GIS
Site: apollo.ogis.state.me.us/
E-mail: http://apollo.ogis.state.me.us/asp/feedinput.htm [feedback form]
Phone: 207-624-8800
Node: No; designation of Maine Office of GIS as an NSDI Clearinghouse is pending.
Data: State agency and federal (DOQ, DEM).
Notes: Site offers wide variety of state and federal data. Most USGS digital data are provided. Data distributed only in Arc/Info format. No charge for data; charges only apply for custom work. Extensive data search capability. Search by region, USGS quad, county, or city. Data provided in USGS quad units. Site supports an active integrated transportation and emergency (E911) database. Interactive map-based access project is planned.

Maryland

Name: University of Maryland, Baltimore County Spatial Analysis Laboratory (UMBC-SAL)
Site: baltimore.umbc.edu/mdnsdi/
E-mail: swalke2@umbc.edu
Phone: 410-455-8020
Node: UMBC-SAL is an NSDI node.
Data: Municipal, county, state, and federal datasets at a variety of scales.
Notes: The site uses an Internet map server to allow users to browse through the site's holdings. Holdings include statewide and nationwide coverages, but with an emphasis on local and municipal data. Large number of data available for the City of Baltimore. Users can select data by theme, source, or region. Also see the Maryland State Government Geographic Information Coordinating Committee (www.msgic.state.md.us/).

Massachusetts

Name: Massachusetts Geographic Information System (MassGIS)
Site: www.state.ma.us/mgis/massgis.htm
E-mail: jane.pfister@state.ma.us
Phone: 617-727-5227 x323
Node: No.
Data: Large variety of themes; USGS products; also includes 1:5000 scale orthophotos.
Notes: This very comprehensive site serves as the general digital data repository for the state. Users benefit from the state's small size, which results in an abundance of large-scale datasets. Some prepackaged data are available on CD for a nominal charge. Other data can be downloaded free from the site. Also see the excellent Internet Geo-Image Library for the Massachusetts Coast site managed by MIT (coast.mit.edu/).

Michigan

Name: Improving Michigan's Access to Geographic Information Networks (IMAGIN)
Site: www.imagin.org/
E-mail: info@imagin.org
Phone: 248-489-3972
Node: No.
Data: FTP site.

Notes: IMAGIN operates an FTP site with sections for free and members-only downloads. See the FTP readme page for the latest information (ftp://ftp.imagin.org/readme.txt).

Minnesota
Name: Minnesota Department of Natural Resources
Site1: deli.dnr.state.mn.us/
Name2: Land Management Information Center (LMIC)
Site2: www.lmic.state.mn.us/
E-mail: robert.maki@dnr.state.mn.us (DNR); lmic@mnplan.state.mn.us (LMIC)
Phone: 651-297-2329 (DNR); 651-296-1211
Node: Both sites are NSDI Clearinghouse nodes.
Data: Wide variety of state and federal products (DOQs, DRGs, and DEMs). Many available at 1:12,000 or greater scale.
Notes: The Minnesota GIS Data Deli serves up natural resource data statewide. The site allows searches by theme and area and also supports interactive (Internet-based) mapping. The LMIC site is a more general state data source. LMIC has the latest information on the status of base mapping and land-use data. The LMIC Minnesota Geographic Data Catalog provides over 100 data sets, many of which are available online. The wide variety of data themes were developed by state and federal agencies. The Minnesota GIS/LIS Consortium has a page with many links to Minnesota data sources (www.mngislis.org/contacts.htm). Also visit the Minnesota Geographic Metadata Guidelines (www.lmic.state.mn.us/gc/stds/metadata.htm).

Mississippi
Name: Mississippi Automated Resource Information System (MARIS)
Site: www.maris.state.ms.us
E-mail: steve@supernova.ihl.state.ms.us
Phone: 601-432-6149
Node: No.
Data: State and federal datasets; socioeconomic, land cover, political boundaries.
Notes: MARIS provides statewide GIS services for Mississippi. Data are available by county and include ten basic layers (roads and Streets, census places, hydrography, boundaries, railroads). The site also has a wealth of information on forestry and land cover due to the forestry resources of the state.

Missouri
Name: Missouri Spatial Data Information Service (MSDIS)
Site: msdis.missouri.edu/

E-mail: msdis@missouri.edu
Phone: 573-884-7802
Node: MSDIS in an NSDI node.
Data: State, federal themes including boundaries, census, flood maps, hydrography, and transportation.
Notes: MSDIS has easy access to numerous statewide data layers (including DOQs). Many are provided by federal sources, such as Census, FEMA, USFS, USGS, and others. Data available through several FTP sites. An Internet Map Server site is being set up for distribution of DOQs and other large datasets.

Montana
Name: Montana Natural Resource Information System (NRIS)
Site: nris.state.mt.us/
E-mail: nris.state.mt.us/comment.html [comment page]
Phone: 406-444-5354
Node: NRIS is an NSDI node.
Data: Regional, state, and federal data. Scales from 1:24,000 and smaller.
Notes: Large variety of data for the state. Many coverages of regional areas where extensive land restoration is underway. Data include many common USGS and census data, particularly vector coverages. The site has lots on GIS-related information on Montana and has served as an early example of efficient statewide data delivery via the Internet.

Nebraska
Name: Nebraska Geospatial Data Clearinghouse
Site: geodata.state.ne.us/
E-mail: lzink@doc.state.ne.us
Phone: 402-471-3206
Node: Yes.
Data: Transportation, geology, boundaries, land cover, many more.
Notes: Also see the University of Nebraska Center for Advanced Land Management Information Technologies (CALMIT) site at www.calmit.unl.edu/calmit.html. The CALMIT site has some USGS data but most holdings are specialized hydrographic and geologic datasets.

Nevada
Name: Nevada Bureau of Mines and Geology (NBMG)
Site: www.nbmg.unr.edu/
E-mail: nbmginfo@unr.edu

Phone: 775-784-6691
Node: Yes.
Data: Geological, printed maps, DRGs, land cover.
Notes: Large FTP site with a number of digital data layers.

New Hampshire
Name: New Hampshire Geographically Referenced Analysis and Information Transfer System (NH GRANIT)
Site: www.granit.sr.unh.edu/
E-mail: granitx@unh.edu
Phone: 603-862-1792
Node: No.
Data: General state and federal framework layers (transportation, hydrography, terrain, etc.).
Notes: GRANIT contains lists of base mapping layers that can be requested or downloaded from the Internet site. Many common USGS layers are available (DEMs, DLGs, DRGs). Most data available at 1:24,000 scale. The site also includes *Landsat* data.

New Jersey
Name: New Jersey Geographic Metadata Clearinghouse
Site: njgeodata.rutgers.edu/
E-mail: agung@njgeodata.rutgers.edu
Phone: 732-932-3822 x727
Node: Yes.
Data: Local, state, and federal data.
Notes: Metadata driven server provides access to local, state, and federal datasets. Searches are conducted through the metadata search gateway at the Federal Geographic Data Committee. Site also allows viewing of maps through the Map Garden webtool.

New Mexico
Name: New Mexico Resource Geographic Information System (RGIS)
Site: rgis.unm.edu/
E-mail: abudge@spock.unm.edu
Phone: 505-277-3622, x231
Node: RGIS is an NSDI node.
Data: Socioeconomic, census, and natural and cultural resource data.

Notes: Metadata driven search tools provide access to information on each data layer. Comprehensive statewide data holdings including many federal (due to land management) and City of Albuquerque datasets. Contact RGIS to obtain data. Cost recovery charges apply. Prepackaged data also available on CD for nominal charge.

New York
Name: New York State GIS Clearinghouse
Site: www.nysl.nysed.gov/gis/index.html
E-mail: nysgis@mail.nysed.gov
Phone: N/A
Node: Yes.
Data: Local, state, and federal framework and socioeconomic datasets.
Notes: Comprehensive site with access to many layers, including cadastral, economic development, emergency response, and political boundaries. Other general framework layers. Additional New York GIS data can be found at the Cornell University Geospatial Information Repository (CUGIR, cugir.mannlib.cornell.edu/).

North Carolina
Name: North Carolina Center for Geographic Information and Analysis (NCGIA)
Site: cgia.cgia.state.nc.us:80/cgia/
E-mail: dataq@cgia.state.nc.us
Phone: 919-733-2090
Node: Yes, North Carolina Geographic Data Clearinghouse.
Data: Framework datasets (DRGs, DOQs, DLGs, DEMs), census data, SSURGO, other natural resource and socioeconomic data.
Notes: The NCGIA site serves as a major source of geographic information for the state. The site is exceptionally detailed, with links to anything GIS in the state. Access to large-scale datasets through the site. NCGIA has links to three sources of data: (1) North Carolina Corporate Geographic Database (NCCGDB), (2) North Carolina Geographic Data Clearinghouse (NCGDC), and (3) North Carolina Geological Survey.

North Dakota
Name: North Dakota's Spatial Data Clearinghouse
Site: www.state.nd.us/ndgs/gis.html

E-mail: ryan@rival.ndgs.state.nd.us
Phone: 701-328-8000
Node: No.
Data: USGS DLGs, DEMs.
Notes: The North Dakota Geological Survey manages this site. The site hosts both large and small-scale USGS digital framework layers. Contact the site for more datasets.

Ohio
Name: Ohio Geographically Referenced Information Program (OGRIP)
Site: www.state.oh.us/das/dcs/ogrip/
E-mail: gis.support@das.state.oh.us
Phone: 614-466-4747
Node: OGRIP is an NSDI node.
Data: USGS (DLGs, DOQs, DRGs, DEMs) and census TIGER.
Notes: Comprehensive site with GIS resources for data, events, contacts, and services. USGS and census products available for download. Data also available on CD for charge. Contact the OGRIP GIS Support Center for more information. Also see the Ohioview site (www.ohioview.org/new-index.html) for access to state satellite imagery.

Oklahoma
Name: Spatial and Environmental Information Clearinghouse (SEIC)
Site: www.seic.okstate.edu/
E-mail: webmaster@seic.lse.okstate.edu
Phone: 405-744-8433
Node: No.
Data: FEMA, USGS (DRGs, DEMs, DLGs, and DOQs).
Notes: Many federal GIS datasets with various scales. Links to weather data through the Oklahoma Mesonet, soils, hydrography. Also see the USDA-NRCS site (www.ok.nrcs.usda.gov/).

Oregon
Name: State Service Center for Geographic Information Systems (SSCGIS)
Site: www.sscgis.state.or.us/
E-mail: gis@web1.css.das.state.or.us
Phone: 503-378-2166
Node: SSCGIS is an NSDI Clearinghouse node.

Data: State agency data, federal datasets (DOQs, DRGs, TIGER, DEMs, and EPA data).
Notes: SSCGIS offers over 100 different themes of data. These data include varying state agency datasets as well as many common USGS digital data at vary scales. The USGS data range from 1:2,000,000 scale DLGs to DOQs. Most data available at 1:24,000 and 1:100,000 scales. Data are available in Arc/Info format and are zipped. SSCGIS also operates as an all-digital ESIC in cooperation with USGS. Data available via Internet, except larger (DOQ) datasets.

Pennsylvania
Name: Pennsylvania Spatial Data Access (PASDA)
Site: www.pasda.psu.edu/
E-mail: www.pasda.psu.edu/comments/comments.html [comment form]
Phone: 814-863-0291
Node: PASDA is an NSDI node.
Data: State and federal (USGS framework data including DOQs).
Notes: Interactive mapping now available through the Pennsylvania Explorer. Site has numerous links to state data sources. FTP site serves variety of framework and other data—many datasets available at 1:24,000. Map gallery.

Rhode Island
Name: Rhode Island Geographic Information System (RIGIS)
Site: www.edc.uri.edu/rigis/
E-mail: maps@uri.edu
Phone: 401-874 5054
Node: No.
Data: Local, state, and federal datasets.
Notes: University of Rhode Island site (http://www.edc.uri.edu/gis/) has access to GIS resources throughout the state, including DOQs, data, projects, and applications. Several versions of each primary framework data layer are available. Additional datasets include cultural, demographic, geologic, land use, and recreation layers.

South Carolina
Name: South Carolina Department of Natural Resources GIS Data Clearinghouse
Site: www.dnr.state.sc.us/gisdata/
E-mail: dbman@water.dnr.state.sc.us

Phone: 803-737-0800
Node: No.
Data: USGS 1:24,000 data.
Notes: Datasets include soils, wetlands, land use, DRGs, DEMs, and DRGs. Also see the University of South Carolina GIS Data Server (www.cla.sc.edu/gis/dataindex.html).

South Dakota
Name: South Dakota Geological Survey (SDGS)
Site: www.sdgs.usd.edu/index.html
E-mail: tcowman@usd.edu
Phone: 605-677-5895
Node: No.
Data: Geological-related data.
Notes: South Dakota GIS resources page. Geological and GPS data.

Tennessee
Name: University of Tennessee Map Library
Site: toltec.lib.utk.edu/~cic/tn.htm
E-mail: maplib@aztec.lib.utk.edu
Phone: N/A
Node: No.
Data: Printed map library.

Texas
Name: Texas Natural Resources Information System (TNRIS)
Site: www.tnris.state.tx.us
E-mail: webmastertnris@tnris.state.tx.us
Phone: 512-463-8337
Node: TNRIS is the primary state site and is an NSDI Clearinghouse node.
Data: DOQs, DLGs, DEMs, DRGs, SSURGO, NWI, and TIGER.
Notes: TNRIS is a division of the Texas Water Development Board and provides analog (paper maps, aerial photos, and census) and digital data to the public. Many GIS datasets available via the Internet, while others need to be supplied on CD. TNRIS provides most standard USGS digital data and is working to create framework datasets statewide. The TNRIS office is open to the public; there is no charge for data, although there are charges for staff and computer time.

Utah

Name: Utah Automated Geographic Reference Center (AGRC)
Site: www.its.state.ut.us/agrc/
E-mail: agrcclark@gis.state.ut.us
Phone: 801-537-9201
Node: No.
Data: State and federal framework datasets from 1:24,000 through 1:100,000 scale.
Notes: In addition to framework data, AGRC has geologic, demographic, ownership, utility, and political data. Data available through FTP site (ftp://ftp.state.ut.us/ftpagrc).

Vermont

Name: Vermont Center for Geographic Information, Inc. (VCGI)
Site: geo-vt.uvm.edu/
E-mail: webmaster@vcgi.uvm.edu
Phone: 802-656-0776
Node: VCGI is an NSDI node.
Data: Large variety of local, state, federal datasets, including many large-scale products.
Notes: Site has access to a huge number of state data. Many datasets are 1:24,000 scale or larger and are available in several formats. Size of state allows for more data variety, new layers, and larger scales. Site has links to programs, including statewide integrated emergency 911 addressing.

Virginia

Name: Geospatial and Statistical Data Center (Geostat)
Site: fisher.lib.virginia.edu/
E-mail: geostat@virginia.edu
Phone: 804-982-2630
Node: No.
Data: Various state resources, some framework data.
Notes: Links to GIS data and interactive mapping page. Site also has links page to other GIS data sources (fisher.lib.virginia.edu/collections/digimaps.html).

Washington

Name: Washington State Geospatial Clearinghouse
Site: metadata.gis.washington.edu/

E-mail: phurvitz@u.washington.edu
Phone: 360-902-3447
Node: Washington State Geospatial Clearinghouse is an NSDI Clearinghouse node.
Notes: Metadata-based search run through the NSDI Clearinghouse search page. A number of Washington groups, including state agencies, university libraries, and several counties, have contributed data along with federal agencies. The Washington Geographic Information Council (www.wa.gov/gic/) has ongoing projects to complete hydrography, transportation, and cadastral data.

West Virginia
Name: West Virginia State GIS Technical Center
Site: wvgis.wvu.edu/
E-mail: wvgis@wvu.edu
Phone: 304-293-5603
Node: Yes.
Data: 1:24,000 DEM, DLG, DOQ
Notes: Black-and-white and color infrared DOQs available for much of state (All DOQs compressed.) Six DLG layers available for various regions. 30m DEMs available statewide. All data available via Internet.

Wisconsin
Name: Wisconsin Land Information Clearinghouse
Site: badger.state.wi.us/agencies/wlib/sco/pages/wisclinc.html
E-mail: sco@facstaff.wisc.edu
Phone: 608-262-3065
Node: WISCLINC is an NSDI Clearinghouse node.
Data: Major Framework themes available by county and statewide.
Notes: Wisconsin has considerable GIS data resources available via the Internet. The WISCLINC site concentrates on numerous themes within Framework (elevations, boundaries and districts, transportation, etc.) and others, such as soils. The Wisconsin Initiative for Statewide Cooperation on Land Analysis and Data (WISCLAND) provides data on land cover, vegetation, wetlands, hydrography, and DRGs (http://feature.geography.wisc.edu/sco/wiscland/wiscland.html).

Wyoming
Name: Wyoming Spatial Data Clearinghouse

Site: wgiac.state.wy.us/wsdc/
E-mail:
Phone: 307-777-5103
Node: WSDC is an NSDI Clearinghouse node.
Data: Major Framework themes available by county and statewide.
Notes: WSDC offers numerous state datasets through their web site. All data are accessible via user searches by theme, region, FGDC Clearinghouse Gateway, federal agency, and scale. Quick links to popular downloads. Information pages on unpacking and using data. The main page for Wyoming GIS information is maintained by the Wyoming Geographic Information Advisory Council (http://wgiac.state.wy.us/index.html).

U.S. Local and Regional Sources

This section provides information on GIS data resources for local and regional government. Only a few sites are listed, but they represent a group with vast geographic resources—in the United States and worldwide. Within the United States alone, there are over 3000 counties, each with its own government and own GIS requirements. The geographic resources attributable to counties range from very large, diverse holdings for large cities down to paper maps only (no digital data) in the most rural counties. This county tally does not, of course, include the many other forms of local and regional government, such as municipalities, metropolitan planning organizations (MPOs), councils of government (COGs), river authorities, and utility districts.

In this appendix I've listed several of the groups that represent counties and their needs. The number of local and regional governmental units is so large that no one site lists all the GIS resources available through these groups. The most effective way to locate data for a particular area is to visit the local and regional sites that cover the area. This method has the added advantage of generally putting you in touch with the most knowledgeable people. Local and regional governments are charged with collecting and maintaining geographic data; state and federal agencies usually provide more general data. *Always* start with the local or regional groups (unless you are very familiar with the geographic data of the area) when collecting data in an unfamiliar region.

Name: National Association of Counties (NACO)
Site: www.naco.org
E-mail: webmaster@naco.org
Phone: 202-393-6226
Notes: NACO represents the interests of county government in the United States. It assists counties with matters ranging from federal legislation to GIS data collection for the Census. NACO realizes the importance of GIS and has

participated actively in nationwide GIS forums and strategy development. It has information on county demographics and has created a GIS Starter Kit to assist counties in using GIS. Contact NACO for more information on getting started with finding and requesting county GIS data.

Name: National Academy of Public Administration (NAPA)
Site: www.napawash.org/napa/index.html
E-mail: cbrouwer@napawash.org
Phone: 202-347-3190
Notes: NAPA is not a GIS data distributor, per se, but it plays an important role in the application of GIS in government. It has conducted studies related to the role of GIS, such as the well-known *Geographic Information for the 21st Century: Building a Strategy for the Nation.*

Name: National League of Cities
Site: www.nlc.org
E-mail: pa@nlc.org
Phone: 202-626-3000
Notes: Group more attuned to serving the needs of elected officials rather than providing geographic information. However, site does have links to related local, regional, and technological groups and societies.

Name: National Association of Regional Councils (NARC)
Site: www.narc.org
E-mail: stephen@narc.org
Phone: 202-457-0710
Capsule: NARC serves the needs of regional governmental groups, typically consisting of a related group of counties. Regional government serves as an excellent source of GIS information. Also see the related site for the Association of Metropolitan Planning Organizations (AMPO, www.narc.org/ampo/index.html), serving metropolitan planning organizations (MPOs).

Name: GeoCommunity
Site: search.geocomm.com/
E-mail: questions@geocomm.com
Phone: 850-897-0110
Notes: The GeoCommunity is a "GeoSpecific Search Engine" that specializes in retrieving difficult to find geographic information sites. This ability is par-

ticularly useful in locating local and regional data sources. For any given locale, there can be a number of local and regional entities that collect information about that site. Having a search engine that can be selective in choosing possible sites will help eliminate multiple hits for certain sites while ignoring others.

Name: ESRI State and Local Government
Site: www.esri.com/industries/localgov/devsrvcs.html
E-mail: web_reply@fws.gov
Phone: 202-208-3100
Notes: The ESRI Internet site has some pages devoted to different applications in local government. The site lists how GIS plays a role locally (infrastructure management, engineering, planning, utilities) and has links to further descriptions of land use and community development, public works, land records, and utility applications.

Name: American Planning Association (APA)
Site: www.planning.org/
E-mail: Library@planning.org
Phone: 312-431-9100
Notes: Of all the organizations representing and assisting local and regional government, the APA has strong ties to GIS because of the use of the technology by planners. While many municipal and local government departments, such as engineering and utilities, use GIS, you can usually rely on finding GIS in city planning. The APA provides a wealth of information about the profession and the goals of planners—goals that can be assisted by GIS. While GIS is becoming a tool that many technologies and professions embrace, the planning field as a whole is very closely tied to GIS.

Private Sources

Name: GIS Data Depot
Site: www.gisdatadepot.com/
E-mail: questions@geocomm.com
Notes: The Data Depot is arguably the largest private source of GIS data in the United States. Most of its datasets are digital USGS products that have been produced throughout the country and are now stored in one place. It could be described as a private version of the USGS EROS Data Center (without the latter's satellite and aerial photography collections) due to its holdings and USGS-spec data. Data are available free of charge when downloaded and custom products can be created on request.

Name: TopoZone
Site: www.topozone.com/
E-mail: info@topozone.com
Notes: TopoZone specializes in the seamless creation and presentation of scanned USGS quadrangle maps (DRGs). Their database includes every standard USGS quadrangle map and is accessible to the public at no charge.

Name: Environmental Systems Research Institute (ESRI)
Site: www.esri.com/data/index
Notes: ESRI offers several sources of GIS data through its Internet site. The Data Hound (gis.esri.com/datahound/main.cfm) is a service that catalogs and searches Internet sites offering free downloadable data. An excellent tool for all GIS users. ESRI also offers other prepackaged data and links to datasets compatible with its ArcView and ArcInfo software.

Name: Cartica
Site: www.cartica.com/cartica/
Notes: Provides search tools to select region and digital products to purchase.
Name: GeoData
Site: www.geodatas.com
Notes: GeoData specializes in digital elevation data (both DEMs and DTMs). Coverages available for the United States and worldwide.

Foreign Sources

Name: World Health Organization (WHO)
Site: www.who.int/ctd/html/hmapwhatis.html

Name: Australian Geological Survey Organisation
Site: www.agso.gov.au/

Name: Australian Surveying and Land Information Group (AUSLIG)
Site: www.auslig.gov.au/
What it is: Australia's national mapping agency.

Name: Western Australian Land Information System (WALIS)
Site: www.walis.wa.gov.au/
What it is: Mapping consortium for the state of Western Australia.

Name: Australia New Zealand Land Information Council
Site: www.anzlic.org.au/

Name: Natural Resources Canada (NRCan)
Site: www.NRcan.gc.ca/

Name: Instituto Nacional de Estadistica, Geografia, e Informatica (INEGI)
Site: www.inegi.gob.mx/difusion/ingles/acercainegi/fiacerine.html
E-mail: atencion.usuarios@inegi.gob.mx
What it is: Mexico's national agency for maps and statistical information.

Name: Swiss Federal Research Institute WSL
Site: www.wsl.ch/welcome-en.ehtml

Name: British Geological Survey (BGS)
Site: www.bgs.ac.uk/

Name: Ordnance Survey
Site: www.ordsvy.gov.uk/
Note: Britain's national mapping agency.

Name: The Federation of Swedish County Councils
Site: www.lf.se/lfenglish

Name: Association for Geographic Information (AGI)
Site: www.agi.org.uk

Name: European Umbrella Organisation for Geographic Information
Site: www.eurogi.org

Name: National System for Geographic Information (SNIG)
Site: snig.cnig.pt/snig/english

Name: Spatial Information Directory
Site: 193.58.158.196/metadata/

Name: Slovenian National Spatial Data Catalogue
Site: www.sigov.si:81/GISborza/MPBeng/Index.html

Name: The SIGNET Project
Site: www.signet.tm.fr/

Name: National Clearinghouse Geo Informatie
Site: www.ncgi.nl/home.html

Name: National Land Survey of Finland
Site: www.kartta.nls.fi

Name: Infodatabase of Geodata
Site: www.geodata-info.dk

Name: Conseil National de l'Information Géographique
Site: www.cnig.fr/

Name: Austrian Umbrella Organisation for Geographic Information (AGEO)
Site: www.ageo.at/

Other GIS Sites of Interest

Thus far, the Appendixes have listed locations of GIS data sites by the major categories of providers. As with many classification exercises, however, there are always some individuals that cannot be placed within a group. This section is for those sites. Some of the sites listed in this appendix will provide GIS data, and some will not. All the sites are valuable in some manner because they provide information that makes for a well-rounded knowledge of geographic data. The sites include professional associations, publications, coordinating bodies, and conference sites.

These groups not only provide more information on locating data, they also help convey the real changes affecting the field of GIS. Most GIS data distribution sites, though they are always updating their data holdings, are relatively static. Constantly changing sites built around very large spatial databases is not something to do often. But the world of geographic thought, technology, and policies is constantly changing our field. These sites can help provide information on trends that will affect your data choices and access in the future.

One interesting sideline to searching for, and accessing, geographic data is that such searches invariably lead to other thoughts. This is particularly true of searching through the Internet. One moment you are looking for orthophoto quads for a particular region. You come across a link to a detailed frequently asked questions (FAQ) section on orthophotos and go there. Perhaps that site leads to information at the USGS on their policy for orthophoto development and plans for designing new base map layers for the country. If you like their planning and ideas, perhaps you would like a position with that agency? A convenient link shows what is currently available. You access that site and then see a kink to . . .

We can see where this is going. You may enter a task with a specific goal but easily be led astray into other related (or even unrelated) subjects. That is what this section is for. While you are searching for GIS data you may start thinking about the following:

- Where can I find other lists and directories listing GIS resources?
- What is the latest news affecting the GIS industry?
- Where can I find information on jobs available in GIS?
- What are some professional associations in GIS that would be beneficial to join?
- Where can I go to get training in the latest GIS software or even find an academic program leading to an advanced degree in geography?

This section contains some links that might help answer these questions. The problem is not how to get started in your GIS-related searching, it's knowing when to stop!

Name: *Directions Magazine*
Site: www.directionsmag.com/
E-mail: website@directionsmag.com
What it is: Directions is an online GIS magazine.
Notes: Directions is an online news source for information on the geographic services industry. It has everything you would expect a trade magazine to have, including the latest industry news, software availability and reviews, links to data, software tools, and GIS books, interviews, discussions on applications and GIS operations, and even letters to the editor.

Name: GEOPlace.com
Site: www.geoplace.com/
E-mail: geomgr@geoplace.com
Phone: 847-427-9512
What it is: Online GIS resources and home to GIS magazines.
Notes: GeoPlace is operated by the Adams Business Media Company and hosts a vast selection of GIS resources and links. GeoPlace is possibly best known for its GIS business magazines, including *GeoWorld* (formerly *GIS World*) and *Business Geographics*. The site lists a number of features with more information, such as industry news, job links, GIS books, events, newsletters, and much more. GeoPlace also hosts the Geodirectory, which provides links to GIS data and service providers by region.

Name: GISLinx.com
Site: www.gislinx.com/

E-mail: info@gislinx.com
Phone: 602-542-4709
What it is: Comprehensive listing of GIS sources.
Notes: Looking for the latest information on GIS links on the Internet? The GISLinx site has links to data, government agencies producing GIS data, GIS education, jobs, software—you name it. The site states that it has over 1700 links to GIS-related materials. The site does not contain information per se, but rather acts as an organized index of connections to other sites. GISLinx is operated by GISLinx Technology Company.

Name: The Association of American Geographers
Site: www.aag.org/
E-mail: gaia@aag.org
Phone: 202-234-1450
What it is: Professional association for the geography profession (not just GIS).
Notes: From the AAG web site: "The Association of American Geographers (AAG) is a scientific and educational society founded in 1904. Its 6,500 members share interests in the theory, methods, and practice of Geography." The AAG sponsors an annual conference and has specialty groups for different interests and practices of geography. The site has links to events, jobs in geography, publications, education, and the AAG specialty groups. AAG is primarily an academic organization but membership is open to students and professionals in the varied applications of geography.

Name: Geospatial Information & Technology Association (GITA)
Site: www.gita.org/
E-mail: staff@gita.org
Phone: 303-337-0513
What it is: Professional organization serving technical applications of GIS.
Notes: GITA, formerly known as Automated Mapping and Facilities Management (AM/FM), specializes in the application of geographic information and information processes in technological applications. GITA sponsors an annual conference as well as conferences for specialized GIS applications and electric utilities.

Name: American Society Photogrammetry and Remote Sensing (ASPRS)
Site: www.asprs.org/
E-mail: asprs@asprs.org

Phone: 301-493-0290
What it is: Society for remote-sensing professionals.
Notes: ASPRS represents the remote-sensing and photogrammetry fields. Members of ASPRS receive the organization journal *Photogrammetric Engineering & Remote Sensing.*

Name: American Congress on Surveying and Mapping (ACSM)
Site: www.survmap.org/
E-mail: infoacsm@mindspring.com
Phone: 301-493-0200
What it is: Professional association for mapping professionals and surveyors.

Name: Urban and Regional Information Systems Association (URISA)
Site: www.urisa.org/
E-mail: info@urisa.org
Phone: 847-824-6300
What it is: Professional association with significant ties to GIS uses in public works and local and state government.

Name: The Canadian Association of Geographers (CAG)
Site: www.uwindsor.ca/cag/
What it is: Professional association of Canadian geographers.
Notes: This site has excellent links to other geographical societies worldwide.

Name: U.S. Census—FAQ
Site: www.census.gov/ftp/pub/geo/www/faq-index.html
Notes: An excellent site with answers and information on GIS that go far beyond Census issues. Information is available on projections, GIS listservers, journals, software, data formats, academic programs, and links to data sources.

Name: Geography Employment Resources
Site: www.cla.sc.edu/GEOG/misc/job.html
Notes: This is an excellent source for GIS job-related sites operated by companies, universities, and associations for jobs both in the United States and abroad.

Name: GIS Jobs Clearinghouse
Site: www.gjc.org/
E-mail: gjc-info@gjc.org

Notes: The oldest and most visited GIS jobs site on the Internet.

Name: GISjobs.com
Site: www.gisjobs.com/
E-mail: info@gisjobs.com
Notes: Site allows users to subscribe to the job-posting mailing list to automatically receive notifications of new job postings matching the user's needs. Excellent jobs resource page with links to companies, universities, and general GIS resources.

Name: National Geographic
Site: www.nationalgeographic.com/maps/index.html
Notes: One of the most popular mapping sites on the Internet. National Geographic has access to their traditional printed atlas maps, a dynamic map server, and lots of general information on mapping and exploration. One of the best sites for general earth science knowledge and archival information, including maps on CD.

INDEX